6543

FOR REVIEW

S. EJAZ AHMED
BOOK REVIEW EDITOR
Technometrics

Cooperative Management

Series editors

Constantin Zopounidis, Chania, Greece
George Baourakis, Chania, Greece

More information about this series at http://www.springer.com/series/11891

Zacharoula Andreopoulou
Christiana Koliouska · Constantin Zopounidis

Multicriteria and Clustering

Classification Techniques in Agrifood
and Environment

Zacharoula Andreopoulou
Forest Informatics Laboratory
Aristotle University of Thessaloniki
Thessaloniki
Greece

Constantin Zopounidis
School of PEM
Technical University of Crete
Chania
Greece

Christiana Koliouska
Forest Informatics Laboratory
Aristotle University of Thessaloniki
Thessaloniki
Greece

ISSN 2364-401X   ISSN 2364-4028 (electronic)
Cooperative Management
ISBN 978-3-319-55564-5   ISBN 978-3-319-55565-2 (eBook)
DOI 10.1007/978-3-319-55565-2

Library of Congress Control Number: 2017934867

© Springer International Publishing AG 2017
This work is subject to copyright. All rights are reserved by the Publisher, whether the whole or part of the material is concerned, specifically the rights of translation, reprinting, reuse of illustrations, recitation, broadcasting, reproduction on microfilms or in any other physical way, and transmission or information storage and retrieval, electronic adaptation, computer software, or by similar or dissimilar methodology now known or hereafter developed.
The use of general descriptive names, registered names, trademarks, service marks, etc. in this publication does not imply, even in the absence of a specific statement, that such names are exempt from the relevant protective laws and regulations and therefore free for general use.
The publisher, the authors and the editors are safe to assume that the advice and information in this book are believed to be true and accurate at the date of publication. Neither the publisher nor the authors or the editors give a warranty, express or implied, with respect to the material contained herein or for any errors or omissions that may have been made. The publisher remains neutral with regard to jurisdictional claims in published maps and institutional affiliations.

Printed on acid-free paper

This Springer imprint is published by Springer Nature
The registered company is Springer International Publishing AG
The registered company address is: Gewerbestrasse 11, 6330 Cham, Switzerland

Preface

Operational Research (OR) combines tools from different disciplines and generates a new set of knowledge for decision making. The broad applicability of its main topics places OR at the heart of many crucial current problems. Therefore, OR entails the construction and application of new quantitative and qualitative modeling methods to real management and economics problems. OR analysts use a range of mathematical methodologies to solve such management decision problems.

Multicriteria Decision Analysis (MCDA) is rooted in OR and it constitutes a broad term and it describes a collection of various methods and tactics explored when multiple criteria should be taken into consideration aiming to result in a decision for a group or for an individual. This book constitutes the first handbook for students to cover multicriteria analysis for total ranking and clustering classification techniques and their application in the agri-environmental sector. It is the only book dealing with the use of OR results as it brings together a group of examples on the much debated issue of decision-making process of enterprises in the agrifood and environmental sector.

Agricultural and environmental decisions will classically comprise multiple criteria, objective and subjective, such as entrepreneurial, biological and social criteria, all important in these decisions. MCDA provided in the book deliver a proper, numerical means of assessing and evaluating those agricultural and environmental decisions taking into account all available parameters. The identification of the highest priority case, as presented in total ranking method of PROMETHEE II and the various types of typology creation through classification with cluster analysis techniques are evaluated among the most important of MCDA and they were selected to be thoroughly presented.

The book is divided into four chapters, as an introduction, presentation of methodologies, detail presentation of whole case studies and finally a concluding chapter.

Chapter 1, "Introduction", provides an introduction to the operational research, focuses on the need for analysis and the ways of use of research results. Operational research intermixes theories and methodologies from mathematics, management science, computer science, operations management, decision support and many more. Multicriteria decision aid, an advanced field of operations research, provides decision-makers and analysts a wide range of methodologies, which are well suited to the complexity of decision problems in various fields. So, the main objective of many classification methodologies is to develop "optimal" classification models, where the term optimal is often restricted to the statistical description of the alternatives, or to the classification accuracy of the developed model given a training sample. Furthermore, the particularities of operational research in the agrifood and environment sector are referred.

Chapter 2, "Methodologies", presents the statistical techniques in broad conceptual terms. It provides a complete coverage of the most common multicriteria ranking technique, PROMETHEE II method and two of the most common cluster generation methods, hierarchical cluster analysis and K-means analysis, are explored. The PROMETHEE II method is part of the outranking relations theory and is based on the construction of an outranking relation that is used to compare the alternatives with some reference profiles characterizing each class. Hierarchical cluster analysis produces a unique set of nested categories or clusters by sequentially pairing variables, clusters, or variables and clusters. K-means cluster analysis is an example of a nonhierarchical cluster analysis method, where k is equal to the number of clusters the researcher wishes to impose upon the data. Additionally, the combination of the two methods for ranking and classification is presented.

Chapter 3, "Applications in Various Agricultural, Food and Environmental Issues" presents and discusses ten examples in order to understand how operational research is used in the agricultural, food and environment sector. PROMETHEE II method is used for ranking skiing centers, agrotourism enterprises and aquaculture units according to certain characteristics. Hierarchical cluster analysis is implemented for the classification of wood enterprises and enterprises that promote renewable energy sources, while K-means analysis is applied for the classification of government agencies in national parks, the typology of prefectures according to agricultural exploitations and the classification of agrifood entities. Finally, the combination of PROMETHEE II method and clustering is applied for ranking and classification of enterprises that promote nature activities in national parks and of enterprises that promote rural production. Basic guidelines for interpreting the results are included to clarify further the methodological concepts.

Finally, Sect. 3.5, "Concluding Remarks and Future Research" contains discussions, comparisons and necessary recommendations for their use. Each of the presented examples assists in gaining a thorough understanding of both statistical and applied issues underlying these techniques. Potential applications of these techniques in problem solving and decision making in agrifood and environment sector research are numerous. The use of these techniques will continue to grow as increased familiarity with the benefits of cases ranking and classification is gained by researchers.

We are confident that the book will be a useful aid for scientists and decision-makers in the agricultural and environmental sector while reading the book will stimulate a fruitful discussion within scientists and experts and will enhance the employment of the methods as well. We trust the book will advance the methods described in new directions and resolutions with both theoretical and practical insights and applications.

Thessaloniki, Greece Zacharoula Andreopoulou
Thessaloniki, Greece Christiana Koliouska
Chania, Greece Constantin Zopounidis

Acknowledgements

We would like to thank our families and the publisher's personnel for their support and assistance in all phases of this endeavor.

Zacharoula Andreopoulou
Christiana Koliouska
Constantin Zopounidis

Contents

1 **Introduction** 1
  1.1 Need for Operational Research 1
  1.2 Use of Research Results 4
  1.3 Decision-Making in the Agrifood Sector 5
  References 7

2 **Methodologies** 13
  2.1 Ranking Methods 13
      2.1.1 PROMETHEE II Method 15
  2.2 Cluster Generation Methods 20
      2.2.1 Hierarchical Cluster Analysis 22
      2.2.2 K-Means Analysis 25
  References 27

3 **Applications in Various Agricultural, Food and Environmental Issues** 33
  3.1 PROMETHEE II Method 33
      3.1.1 Ranking of Skiing Centers 33
      3.1.2 Ranking of Agrotourism Enterprises 36
      3.1.3 Ranking of Aquaculture Units 38
  3.2 Hierarchical Cluster Analysis 42
      3.2.1 Classification of Wood Enterprises 42
      3.2.2 Typology of RES Enterprises 46
  3.3 K-Means Analysis 52
      3.3.1 Classification of Government Agencies in National Parks 53
      3.3.2 Typology of Prefectures According to Agricultural Exploitations 56
      3.3.3 Classification of Agrifood Entities 63
  3.4 Combining PROMETHEE II and Clustering for Decision Making 65

      3.4.1 Ranking and Classification of Enterprises that Promote
Nature Activities in National Parks 66
      3.4.2 Ranking and Classification of Enterprises that Promote
Rural Production 74
3.5 Concluding Remarks and Future Research................... 78
References 79

Chapter 1
Introduction

1.1 Need for Operational Research

Operations research consists an interdisciplinary field concerning the application of advanced analytical methods to help and improve the decision-making process. More specifically, operations research intermixes theories and methodologies from mathematics, management science, computer science, operations management, decision support and many more. The broad applicability of its main topics places operations research at the heart of many crucial current problems such as economics, psychology, ergonomics, knowledge management, education, quality management, biology, communication network management, supply-chain management, pricing and revenue management, market design, bio-informatics, production scheduling, energy and environmental policy, agricultural planning, data analysis, distribution of goods and resources, emergency and rescue operations, engineering systems design, environmental management, health care management, inventory control, manpower and resource allocation, manufacturing of goods, military operations, risk management, sequencing and scheduling of tasks and traffic control transportation logistics. Therefore, operations research entails the construction and application of new quantitative and qualitative modeling methods to real management and economics problems.

The technological advances in combination with the establishment of operations research as a sound approach to decision making problems, created a new context for addressing real-world problems through integrated, flexible and realistic methodological approaches (Doumpos and Zopounidis 2002). Participatory decision-making is a time-consuming and resource-intensive process (Keseru et al. 2016). During the first decades, operational research has been based on the concept of optimization (Teghem et al. 1989). Operational research analysts collect, synthesize, and work with information, while they use information systems as a source of data and a means of implementing solutions.

Operation research analysts use a range of mathematical methodologies to solve such management decision problems. Firstly, operational research analysts represent the system mathematically and they develop an algebraic or computational model of the system. Then, they use computers to solve the model and finally, to conclude with the best decisions. They build models and modify constraints or variables. However, the applicable methodologies require regularly improvements and expansion in order to accurately represent and forecast current relationships.

The analytical methods that analysts use, include decision analysis, system analysis, management science, control theory, optimization theory, artificial intelligence, fuzzy decision-making, mathematical logic, simulation, network analysis, queuing theory, game theory multi-criteria analysis. Operation research applies different approaches to different types of problems:

- dynamic programming, linear programming, and critical path method are used in handling complex information in allocation of resources, inventory control, and in determining economic reorder quantity.
- forecasting and simulation techniques such as Monte Carlo method are used in situations of high uncertainty such as market trends, next period's sales revenue, and traffic patterns.

The following three steps can describe the process:

- A set of possible alternative solutions to the given problem is developed.
- The alternative solutions are analyzed and reduced to a smaller set of solutions.
- The remaining alternative solutions are subjected to simulated implementation.

Doumpos et al. (2016) provides a unique overview of the fundamental aspects of robustness in OR/MS and the state-of-the-art advances in the related research, adopting a broad perspective that covers different established and emerging OR/MS fields, namely decision aiding, optimization, and analytics.

Multicriteria Analysis (MA) consists the background of the Multicriteria Decision Support Systems (MCDSSs) which includes models, methods and approaches that aim to aid the Decision Makers (DMs) to handle semistructured decision problems with multiple criteria, where the components are transitional and the required information insufficient (Siskos and Spyridakos 1999). Multicriteria decision making (MCDM) refers to making preference decision (e.g., evaluation, prioritization, and selection) over the available alternatives that are characterized by multiple, usually conflicting, criteria (Ma et al. 2010).

Since the early 1970s, MCDM techniques have been developed into many forms and been extensively employed for a wide range of different case studies, such as river basin planning and groundwater remediation (Moeffaert 2003; Weng et al. 2010). At this time, many distinct methodologies for multicriteria decision making problems exist, which can be categorized in a variety of ways, such as form of model (e.g. linear, non-linear, stochastic), characteristics of the decision space (e.g. finite or infinite), or solution process (e.g. prior specification of preferences or interactive) (Gal et al. 2013). Normal MCDM methods have criteria at the same level and deal with subjective data only (Zhang et al. 2009). Last years, multicriteria analysis presents important improvements, which are reflected from the progress of the four theoretical trends (Siskos and Spyridakos 1999):

1. the Value System approaches (American School; Fishburn 1970, 1972, 1982; French 1993; Keeney and Raiffa 1976; Keeney 1992; Von Winterfeldt and Edwards 1993) aiming to the construction of a value system that aggregates the DM's preferences on the criteria based on strict assumptions (complete and transitive preference relation). The estimated value system by this approach provides a quantitative way that leads the DM's in his/her final decision.

2. The French School (Roy 1976, 1985, 1989, 1990; Roy and Bouyssou 1993; Vincke 1992; Brans and Mareschal 1989; Vanderpooten 1990) using a non compensatory approach aims to the construction of a relation (Outranking Relation) that allow the incomparability among the decision actions. The Outranking Relation approach is not bounded into a mathematical model but providing further exploitation and processes deduce to support the DM to conclude to a "good" decision.

3. The Disaggregation-Aggregation approach (Jacquet-Lagreze 1984; Jacquet-Lagreze and Siskos 1982; Siskos 1980; Siskos and Yannacopoulos 1985; Siskos et al. 1993; Pardalos et al. 2013), which consists the third theoretical trend, aim to analyze the DM behavior and cognitive style. Special iterative interactive procedures are used, where the components of the problem and DM's global judgement policy are analysed and following are aggregating into a value system. The target of this approach is to aid the DM to improve his/her knowledge on the problem's state and his/her way of preferring that entail a consistent decision to be achieved.

4. The fourth theoretical trend is the Multiobjective Optimisation approach (Zeleny 1974, 1982; Evans and Steuer 1973; Zionts and Wallenious 1976, 1983; Siskos and Despotis 1989; Korhonen 1990; Jaszkiewicz and Slowinski 1995; Jacquet-Lagreze et al. 1987; Wierzbicki 1992), which consists an extension of the Mathematical Programming one, aiming to solve problems where there are no discrete alternative actions and the objectives are more than one. The solution is estimated through iterative procedures which lead to: (a) achieving the satisfaction levels of the DM on the criteria or; (b) constructing a utility model of the DM that is used for the selection of the solutions that are assessed from a utility maximization procedure; or (c) a combination of the above two described methods.

These methodologies share common characteristics of conflict among criteria, incomparable units, and difficulties in selection of alternatives. The models of MCDM are further divided into multi-objective decision making (MODM) and multi-attribute decision making (MADM) (Climaco 1997). In multiple objective decision making, the alternatives are not predetermined but instead a set of objective functions is optimized subject to a set of constraints. The most satisfactory and efficient solution is sought. In this identified efficient solution it is not possible to improve the performance of any objective without degrading the performance of at least one other objective. In multiple attribute decision-making, a small number of alternatives are to be evaluated against a set of attributes which are often hard to quantify. The best alternative is usually selected by making comparisons between alternatives with respect to each attribute (Pohekar and Ramachandran 2004). The MCDM framework has the advantage of taking into account the specific preference system of any particular investor, while allows for synthesizing in a one and single procedure all the theoretical and practical aspects of the portfolio management (PM) theory (Xidonas et al. 2011).

Multicriteria methods are capable of dealing with the multiple dimensions of evaluation problems (e.g., social, cultural, ecological, technological, institutional, etc.) and give due attention to interest conflicts among stakeholders involved. In general, the aim of these methods is to combine assessment methods with judgement methods and to offer a solid analytical basis for modern decision analysis (Vreeker et al. 2002). MCDM provides a simple but structured framework for controlling the decision-making process while the simplicity of MCDM outputs makes it easier to communicate, coordinate and aggregate individual analyses in the decision-making process (Bui 1987). Jarke (1986) states that MCDM methods can serve as formal tools for preference surfacing, preference aggregation, negotiation, and mediation, both in co-operative and in non-cooperative decision situations.

The characteristic of a multicriteria decision support system (MCDSS) is that it allows the decision maker to interact with the system at any time, setting the parameters of the problem and data to be used (Dimitras et al. 1996). Recent development in computer technology has provided new opportunities to enhance the use of MCDA (Bhargava et al. 2007; Liu and Stewart 2004; Power and Sharda 2007; Shim et al. 2002). For example, with today's multimedia facilities, MCDA methods can be visualized to help preference elicitation and the analysis of the results (Mustajoki and Hamalainen 2007). Increased computational capabilities have also made it possible to create interactive software with new computationally demanding methods, such as linear programming. In recent years, the proliferation of the World Wide Web has enabled MCDSSs that are offered and maintained in one location and are still easily available for remote use (Mustajoki and Hamalainen 2007). An internet-based decision support tool allows stakeholders to explore various management options and discover the individual environmental, economic, and social impacts of each (Zeng and Trauth 2005).

1.2 Use of Research Results

Multicriteria decision aid (MCDA), an advanced field of operations research, DMs and analysts a wide range of methodologies, which are well suited to the complexity of decision problems in various fields (Zopounidis and Doumpos, 2002a, b). So, the main objective of many classification methodologies is to develop "optimal" classification models, where the term optimal is often restricted to the statistical description of the alternatives, or to the classification accuracy of the developed model given a training sample (Doumpos and Zopounidis 2002).

The major characteristic shared by all MCDA classification approaches is their focus on the modeling and addressing of sorting problems. This characteristic of MCDA can be considered as a significant advantage within a decision making context (Doumpos and Zopounidis 2002). The great success of M.A. today is manifested by the increasing number of theoretical and application papers which have been published in scientific journals of operations research and decision science and by the great number of communications presented in scientific meetings (Pardalos et al. 2013).

MCDA can help to ensure transparency during the decision making process (Geldermann et al. 2009; French and Geldermann 2005; Geldermann et al. 2005; Hamalainen et al. 2000; Keefer et al. 2004; Stewart 1992). MCDA methods can be useful for supporting a strategy team tasked with designing and selecting high-value strategic options (Montibeller and Franco 2010). Hallefjord and Jornsten (1986) emphasize that the purpose of applying MCDM to long-term planning problems is to explore solutions, to generate alternative strategies, and to gain insight into the problem, rather than to find optimum solutions (Liu and Stewart 2004).

MCDM provides a framework for group decision and negotiation support that may be oriented around the spaces where individuals can make decisions, in which the decisions can be evaluated and compared by an individual or by the group (Matsatsinis and Samaras 2001). We often need a group of evaluators because many criteria require multiple perspectives of different people, as one evaluator may not have enough knowledge to well assess alone (Zhang et al. 2009). Therefore, group decision-making (GDM) is appropriate to be applied for arriving at a judgment based on the input and feedback of multiple individuals (Bose et al. 1997; Lu et al. 2007; Yager 1993).

1.3 Decision-Making in the Agrifood Sector

When researchers have to solve problems in the agrifood sector, they usually select outranking methods for multiple criteria decision analysis and especially, the PROMETHEE II methodology to perform evaluation and ranking tasks, for the following reasons: (a) because the estimated relation of superiority (of one website over another) is less sensitive in small changes and that offers an easier analysis and discussion of the results (Zopounidis 2001), (b) the use of the superiority relation in the PROMETHEE method is applied when the alternative solutions (websites) have to be ranked from the best to the worst (Zopounidis 2001), and (c) the procedure of assessing and ranking complicated cases of websites is proper for the application of the above methodology in the sense that it is closer to reality (Zopounidis 2001). In fact, there exist two types of the PROMETHEE methodology, the PROMETHEE I that ranks partially and also, the PROMETHEE II, which performs a full and complete ranking, based on all of the input data. The PROMETHEE II methodology was applied in this project because an overall ranking was required. It is also important that our variables concern qualitative data and PROMETHEE II methodology can successfully deal with that prerequisite (Koutroumanidis et al. 2004).

Moreover, regarding the application of PROMETTHEE II in the field of agriculture and environment, there are recent research papers in Greece where the method is successfully applied (Koutroumanidis, et al. 2002; Andreopoulou et al. 2005, Polyzos and Arabatzis 2006; Andreopoulou et al. 2008, 2009).

The PROMETHEE methodology fits better to the targets of the projects in the agrifood sector even if it is compared to other well-established methods. For example, the ELECTRE methods are methods of superiority that use the rule of majority inside a relation of superiority. The target in the ELECTRE is to determine an alternative website, which is relatively "good", based on a majority of criteria without been too "bad" according to the rest of the criteria (Koutroumanidis et al. 2004). Nevertheless, this is not the objective of this project where the objective is the total evaluation of the websites. The AHP method is also well-known and broadly applied (Koutroumanidis, et al. 2004). But, according to Alphonce (1997) the ability of the AHP to analyze different decision factors without the need for a common numerate, other than the decision maker's assessments, makes it one of the favorable multicriteria decision support tools when dealing with complex socioeconomic problems in developing countries (Andreopoulou et al. 2009; Tsekouropoulos et al. 2013).

A theoretical description of a methodology aiming to optimize websites of rural production sectors is described by Andreopoulou et al. (2011) using PROMETHEE II ranking. The reason for application of PROMETHEE method lies in certain advantages of this method compared to used PCA method, which is reflected in the way the problems in this sector could be structured and explored (Arsic et al. 2012). Regarding the application of PROMETHEE II and cluster analysis in the field of agriculture, food and environment, there are recent research papers in Greece where the methods are successfully applied for local development agencies and regional development (Andreopoulou et al. 2007a, b; Arabatzis et al. 2010), forest issues/environment (Andreopoulou 1999; Andreopoulou and Iliadis 2003a, b; Andreopoulou, Koutroumanidis and Manos 2009; Andreopoulou and Papastavrou 2005; Koliouska and Andreopoulou 2011; Kokkinakis et al. 2006), aquacultute (Andreopoulou et al. 2009; Andreopoulou et al. 2006; Kokkinakis and Andreopoulou 2008; Lemonakis et al. 2015), agrifood (Andreopoulou et al. 2008; Tsekouropoulos et al. 2012d, 2013a, b, 2014), regional development/tourism/skiing centers (Andreopoulou and Koutroumanidis 2009; Andreopoulou et al. 2012, 2014a; Tsekouropoulos et al. 2012b; Zopounidis et al. 2014), environmental organizations (Andreopoulou et al., 2010; Andreopoulou et al. 2011), rural production (Andreopoulou et al. 2011, 2014b; Koutroumanidis et al. 2002; Tsekouropoulos et al. 2012c, 2013a, b), Greek prefectures according to tourist resources (Polyzos and Arabatzis, 2006) and National Parks (Andreopoulou et al. 2015; Athanasiadis and Andreopoulou 2010; Koliouska et al. 2017).

References

Alphonce, B. C. (1997). Application of the analytic hierarchy process in agriculture in developing countries. *Agricultural Systems, 53*, 97–112.

Andreopoulou, Z. (1999). The contribution of forest informatics in planning of Forest Administration. Ph.D. Thesis. Aristotle University of Thessaloniki. in greek with english abstract.

Andreopoulou, Z., & Iliadis L. (2003b). The ε-organization of human resources data and its potential in regional Forest Administration in Greece. In *Proceedings of 8th CEST International Conference on Environmental Science and Technology of Global Network for environmental Science and Technology (Global NEST) Organization* (Vol. 2, pp. 31–38) 8–10 Sept 2003. Myrina, Lemnos Island, Greece. (ISSN 1106-5516, ISBN 960-7475-24-0) Proceedings Citation Index.

Andreopoulou, Z., & Iliadis, L. (2003a). Development of a visual decision-support tool for forest service, concerning the classification of forest prefectures according to human resources and forest fire breakouts. In *Proceedings of International Conference of IUFRO "Decision Support for Multiple Purpose Forestry—A Tran Disciplinary Conference on the Development and Application of Decision Support Tools for Forest Management"*. BOKU University of Natural Resources and Applied Life Sciences, 23–25 April 2003, Vienna. Austria, 2003.

Andreopoulou, Z., & Koutroumanidis, T. (2009). Assessment of the ICT adoption stage in eco-agrotourim websites in Greece. In *6th International Conference of Management of Technological Changes* (pp. 441–444), 3–5 September, Alexandroupolis, Greece.

Andreopoulou, Z. S., Vlachopoulou, M., Manos, B., Vassiliadou, S., & Papathanassiou, J. (2005). Website evaluation in the context of support and promotion for e-business in forestry sector. In *Proceedings of International Conference on Information Technology in Agriculture, Food and Environment (ITAFE)* (Vol. I, pp. 353–358), October 2005, Adana, Turkey.

Andreopoulou, Z. S., Kokkinakis, A. K., & Manos, B. (2006). Classification of Greek Fish farming entities and evaluation of their presence in the internet. In *Proceedings of the 3rd International Conference of HAICTA, Hellenic Association of ICT in Agriculture, Food and Environment* 20–23 September 2006, University of Thessaly, Volos, Greece, (SET 960-8029-42-2, ISBN 960-8029-45-7).

Andreopoulou, Z., Arabatzis, G., Manos, B., & Sofios, S. (2007a). Promotion of rural regional development through the WWW. *International Journal of Applied Systems Studies, 1*(3), 290–304.

Andreopoulou, Z. S., Koutroumanidis, T., Arabatzis, G., & Manos, B. (2007b). Ranking of local development practices. Chapter 4. In N. Matsatsinis & C. Zopounidis (Eds.), *Multi-criteria decision system*. Athens: Klidarithmos. ISBN 978-960-461-068-6.

Andreopoulou, Z. S., Tsekouropoulos, G., Koutroumanidis, T., Vlachopoulou, M., & Manos, B. (2008). Typology for e-business activities in the agricultural sector. *International Journal of Business Information Systems, 3*(3), 231–251.

Andreopoulou, Z. S., Kokkinakis, A. K., & Koutroumanidis, T. (2009). Assessment and optimization of e-commerce websites of fish culture sector. *Operational Research, 9*(3), 293–309.

Andreopoulou, Z., Koutroumanidis, T., & Manos, B. (2011). Optimizing collaborative E-commerce websites for rural production using multicriteria analysis. In K. Malik & P. Choudhary (Eds.), *Business organizations and collaborative web: Practices, strategies and patterns* (pp. 102–119). PA, USA: IGI Global.

Andreopoulou, Z., Tsekouropoulos, G., & Pavlidis, Th. (2012). Adoption and perspectives of network technologies and E-marketing in skiing centres websites in the internet. *Journal of Environmental Protection and Ecology, 13*, 2416–2422.

Andreopoulou, Z., Tsekouropoulos, G., Koliouska, C., & Koutroumanidis, T. (2014a). Internet marketing for sustainable development and rural tourism. *International Journal of Business Information Systems, 16*(4), 446–461.

Andreopoulou, Z., Tsekouropoulos, G., Theodoridis, A., Samathrakis, V., & Batzios, C. (2014b). Consulting for sustainable development, information technologies adoption, marketing and entrepreneurship issues in livestock farms. *Procedia Economics and Finance, 9,* 302–309.

Andreopoulou, Z., Koliouska, C., Lemonakis, C., & Zopounidis, C. (2015). National Forest Parks development through Internet technologies for economic perspectives. *Operational Research, 15*(3), 395–421.

Arabatzis, G., Andreopoulou, Z., Koutroumanidis, Th, & Manos, B. (2010). E-government for rural development: Classifying and ranking content characteristics of development agencies websites. *Journal of environmental protection and ecology., 11*(3), 1138–1149.

Arsic, M., Nikolic, D., Mihajlovic, I., Zivkovic, Z., & Djordjevic, P. (2012). Monitoring of ozone concentrations in the Belgrade urban area. *Journal of Environmental protection and ecology, 13*(4), 2057–2067.

Athanasiadis A., & Andreopoulou Z. (2010). Model environmental website promoting forest recreation near wetlands in Greece. Chapter 1. In *Collective Volume of Scientific Papers 3. Innovative Applications of Informatics in Rural Sector and the Environment, HAICTA* (pp. 23–35). ISBN 978-960-357-097-4.

Bhargava, H. K., Power, D. J., & Sun, D. (2007). Progress in Web-based decision support technologies. *Decision Support Systems, 43*(4), 1083–1095.

Bose, U., Davey, A. M., & Olson, D. L. (1997). Multi-attribute utility methods in group decision making, past applications and potential for inclusion in GDSS. *Omega, 25,* 691–706.

Brans, J. P., & Mareschal, B. (1989). The PROMETHEE methods for MCDM, the PROMCALC, GAIA and Bankadviser Software. Vrije Universiteit Brussel, STOO/224.

Bui, T. X. (1987). Co-oP: A group decision support system for cooperative multiple criteria group decision making. Lecture Notes in Computer Science, No. 290. Berlin: Springer.

Climaco, J. (1997). *Multicriteria analysis.* New York: Springer.

Dimitras, A. I., Zanakis, S. H., & Zopounidis, C. (1996). A survey of business failures with an emphasis on prediction methods and industrial applications. *European Journal of Operational Research, 90*(3), 487–513.

Doumpos, M., & Zopounidis, C. (2002). *Multicriteria decision aid classification methods* (Vol. 73). Springer Science & Business Media.

Doumpos, M., Zopounidis, C., & Grigoroudis, E. (2016). *Robustness analysis in decision aiding, optimization, and analytics.*

Evans, J. P., & Steuer, R. E. (1973). A revised simplex method for linear multiple objective programs. *Mathematical Programming, 5*(1), 54–72.

Fishburn, P. C. (1970). *Utility theory for decision making.* New York: Wiley.

Fishburn, P., 1972. Mathematics of decision theory. UNESCO.

Fishburn, P. C. (1982). *The foundation of expected utility.* Dordrecht, Holland: Reidel.

French, S. (1993). *Decision theory: An Introduction to the mathematics of rationality.* West Sussex: Ellis Horwood.

French, S., & Geldermann, J. (2005). The varied contexts of environmental decision problems and their implications for decision support. *Environmental Science & Policy, 8*(4), 378–391.

Gal, T., Stewart, T., & Hanne, T. (2013). *Multicriteria decision making: Advances in MCDM models, algorithms, theory, and applications* (Vol. 21). Springer Science & Business Media.

Geldermann, J., Treitz, M., Bertsch, V., & Rentz, O. (2005). Moderated decision support and countermeasure planning for off-site emergency management. In *Energy and Environment* (pp. 63–80). US: Springer.

Geldermann, J., Bertsch, V., Treitz, M., French, S., Papamichail, K. N., & Hämäläinen, R. P. (2009). Multi-criteria decision support and evaluation of strategies for nuclear remediation management. *Omega, 37*(1), 238–251.

Hallefjord, A., Jornsten, K., & Eriksson, O. (1986). A long range forestry planning problem with multiple objectives. *European Journal of Operational Research, 26*(1), 123–133.

References

Hamalainen, R. P., Lindstedt, M. R., & Sinkko, K. (2000). Multiattribute risk analysis in nuclear emergency management. *Risk Analysis, 20*(4), 455–468.

Jacquet-Lagreze, E. (1984). PREFCALC: Evaluation et decision multicitere. *Revue de l'Utilisateur de l'IBM PC, 3*, 38–55.

Jacquet-Lagreze, E., & Siskos, Y. (1982). Assessing a set of additive utility functions for multicriteria decision making. *European Journal of Operational Research, 10*(2), 151–164.

Jacquet-Lagreze, E., Meziani, R., & Slowinski, R. (1987). MOLP with an interactive assessment of a piecewise-linear utility function. *European Journal of Operational Research, 31*, 350–357.

Jarke, M. (1986). Knowledge sharing and negotiation support in multiperson decision support systems. *Decision Support Systems, 2*, 93–102.

Jaszkiewicz, A., & Slowinski, R. (1995). The light beam search: Outranking based interactive procedure for multiple-objective mathematical programming. In P. M. Pardalos, Y. Siskos & C. Zopounidis (Eds.), *Advances in multicriteria analysis* (pp. 129–146). Dordrecht: Kluwer Academic Publishers.

Keefer, D. L., Kirkwood, C. W., & Corner, J. L. (2004). Perspective on decision analysis applications, 1990–2001. *Decision analysis, 1*(1), 4–22.

Keeney, R., & Rai A. H., 1976. *Decisions with multiple objectives: Preferences and value trades*. New York: Wiley.

Keeney, R. L. (1992). *Value-focused thinking: A path to creative decision making*. London: Harvard University Press.

Keseru, I., Bulckaen, J., & Macharis, C. (2016). The multi-actor multi-criteria analysis in action for sustainable urban mobility decisions: The case of Leuven. *International Journal of Multicriteria Decision Making, 6*(3), 211–236.

Kokkinakis A. K., Andreopoulou Z. S. (2008). Evaluation of the fishery production through classification in the two trans-boundary Prespa lakes. In *Proceedings of the 4th International Conference on Aquaculture, Fisheries Technology & Environmental Management (AQUAMEDIT 2008)*, 21–22/11/2008, Athens, Greece, Abstract Book & CDrom full papers. ISBN: 978-960-87795-2-5.

Kokkinakis, A. K., Andreopoulou, Z. S., & Arabatzis, G. (2006). Typology of Northern Greece rivers aiming to the proper management for the protection of their environmental sensitive fish Fauna. In *Proceedings of the 3rd International Conference of HAICTA, Hellenic Association of ICT in Agriculture, Food and Environment Conference* (pp. 896–902), 20–23 Sept 2006, University of Thessaly, Volos, Greece (ISBN 960-8029-45-7).

Koliouska, C., & Andreopoulou, Z. (2011). Climate change in the Greek Internet. In *15th Pan-Hellenic Forestry Conference, Multi-purpose Forestry and Climatic Change—Protection and Development of Natural Resources*. Karditsa, Greece, 16–19 October. p. 9. (cd in greek).

Koliouska, C., Andreopoulou, Z., Zopounidis, C., & Lemonakis, C. (2017). E-commerce in the context of protected areas development: A managerial perspective under a multi-criteria approach. In *Multiple Criteria decision making* (pp. 99–111). Springer International Publishing.

Korhonen, P. (1990). A multiple objective linear programming decision support system. *Decision Support Systems, 6*, 243–252.

Koutroumanidis, T., Papathanasiou, I., & Manos, B. (2002). A multicriteria analysis of productivity of agricultural regions of Greece. *Operational research: An International Journal, 2*(3), 339–346.

Koutroumanidis, T., Iliadis, L., & Arabatzis, G. (2004). Evaluation and forecasting of the financial performance of the rural cooperatives by a decision support system. *Japanese Journal of Rural Economics, 6*, 31–44 (International Bibliography of the Social Sciences (IBSS) Journal list, Center for Academic Publications Japan (CAPJ).

Lemonakis, C., Andreopoulou, Z., Sfendurakis, I., Lianoudaki, K., & Garefalakis, A. (2015). Prediction of firms financial Distress: The case of the Greek Fish-farming industry. *Journal of Environmental Protection and Ecology, 16*(2), 528–538.

Liu, D., & Stewart, T. J. (2004). Object-oriented decision support system modelling for multicriteria decision making in natural resource management. *Computers & Operations Research, 31*(7), 985–999.

Lu, J., Zhang, G., Ruan, D., & Wu, F. (2007). *Multi-objective group decision making: Methods, software and applications with fuzzy set technology*. London: Imperial College Press.

Ma, J., Lu, J., & Zhang, G. (2010). Decider: A fuzzy multi-criteria group decision support system. *Knowledge-Based Systems, 23*(1), 23–31.

Matsatsinis, N. F., & Samaras, A. P. (2001). MCDA and preference disaggregation in group decision support systems. *European Journal of Operational Research, 130*(2), 414–429.

Moeffaert, D. V. (2003). Multicriteria decision aid in sustainable urban water management. MSc thesis, Industrial Ecology, Royal Institute of Technology.

Montibeller, G., & Franco, A. (2010). Multi-criteria decision analysis for strategic decision making. In *Handbook of multicriteria analysis* (pp. 25–48). Berlin: Springer.

Mustajoki, J., & Hamalainen, R. P. (2007). Smart-Swaps—A decision support system for multicriteria decision analysis with the even swaps method. *Decision Support Systems, 44*(1), 313–325.

Pardalos, P. M., Siskos, Y., & Zopounidis, C. (2013). *Advances in multicriteria analysis* (Vol. 5). Springer Science & Business Media.

Pohekar, S. D., & Ramachandran, M. (2004). Application of multi-criteria decision making to sustainable energy planning—a review. *Renewable and Sustainable Energy Reviews, 8*(4), 365–381.

Polyzos, S., & Arabatzis, G. (2006). Multicriteria approach of the Greek prefectures evaluation according to tourist resources. *Tourism Today*, 96–111.

Power, D. J., & Sharda, R. (2007). Model-driven decision support systems: Concepts and research directions. *Decision Support Systems, 43*(3), 1044–1061.

Roy, B. (1976). From optimization to multicriteria decision aid: Three main operational attitutes. In: H. Thiriez & S. Zionts (Eds.), *Multiple criteria decision making* (Vol. 130, pp. 1–32). Berlin: Springer.

Roy, B. (1985). Methodologie multicrit ere d'aide a la d ecision. Economica, Paris.

Roy, B. (1989). The outranking approach and the foundations of Electre methods. In: C. Bana e Costa (Ed.), *Readings on multiple criteria decision aid* (pp. 155–183). Berlin: Springer.

Roy, B. (1990). Decision aid and decision making. *European Journal of Intelligent Systems in Accounting, Finance and Journal of Operational Research, 45*, 324–331.

Roy, B., & Bouyssou, D. (1993). Aide multicrit ere a la d ecision: M ethodes et cas. Economica, Paris.

Shim, J. P., Warkentin, M., Courtney, J. F., Power, D. J., Sharda, R., & Carlsson, C. (2002). Past, present, and future of decision support technology. *Decision Support Systems, 33*(2), 111–126.

Siskos, Y. (1980). Comment modeliser les preferences au moyen de fonctions d'utilit e additives. *RAIRO Recherche Operationnelle, 14*, 53–82.

Siskos, J., & Despotis, D. K. (1989). A DSS oriented method for multiobjective linear programming problems. *Decision Support Systems, 5*(1), 47–55.

Siskos, Y., & Spyridakos, A. (1999). Intelligent multicriteria decision support: Overview and perspectives. *European Journal of Operational Research, 113*(2), 236–246.

Siskos, Y., & Yannacopoulos, D. (1985). UTASTAR, an ordinal regression method for building additive value functions. *Investigacao Operacional, 5*(1), 39–53.

Siskos, Y., Spyridakos, A., & Yannacopoulos, D. (1993). MINORA: A multicriteria decision aiding system for discrete alternatives. *Journal of Information Science and Technology, 2*(2), 136–149.

Stewart, T. J. (1992). A critical survey on the status of multiple criteria decision making theory and practice. *Omega, 20*(5), 569–586.

References

Teghem, J., Delhaye, C., & Kunsch, P. L. (1989). An interactive decision support system (IDSS) for multicriteria decision aid. *Mathematical and Computer Modelling, 12*(10), 1311–1320.

Tsekouropoulos, G., Andreopoulou, Z., Seretakis, A., Koutroumanidis, T., & Manos, B. (2012a). Optimizing e-marketing criteria for customer communication in food and drink sector in Greece. *International Journal of Business Information Systems., 9*(1), 1–25.

Tsekouropoulos, G., Andreopoulou, Z., Koliouska, C., Koutroumanidis, T., Batzios, C., & Lefakis, P. (2012b). Marketing policies through the internet: The case of skiing centers in Greece. *Scientific Bulletin-Economic Sciences, 11*(1), 66–78.

Tsekouropoulos, G., Andreopoulou, Z., Samathrakis, V., & Grava, F. (2012c). Sustainable Development through agriculture entrepreneurship opportunities: Introducing internet consulting for market places. *Journal of Environmental Protection and Ecology, 13,* 2340–2348.

Tsekouropoulos, G., Andreopoulou, Z., Koliouska, C., Koutroumanidis, T., Batzios, C., & Samathrakis, V. (2013a). Internet functions in marketing: Multicriteria ranking of agricultural SMEs websites in Greece. *Journal of Agricultural Informatics, 4*(2), 22–36.

Tsekouropoulos, G., Koliouska, C., & Andreopoulou, Z. (2013b). Marketing and digital functions in rural agribusiness: A case of classification. *Journal of Marketing Vistas (JMV), 3*(2), 1–10.

Tsekouropoulos, G., Vatis, S.-E., Andreopoulou, Z., Katsonis, N., & Papaioannou, E. (2014). The aspects of internet-based management, marketing, consumer's purchasing behavior and social media towards food sustainability. *Review of Studies in Sustainability, 2014*(2), 207–222.

Vanderpooten, D. (1990). The construction of prescriptions in outranking methods. In *Readings in multiple criteriadecision aid* (pp. 184–215). Berlin Heidelberg: Springer

Vincke, P. (1992). *Multicriteria decision-aid.* West Sussex: Wiley.

Von Winterfeldt, E., & Edwards, W. (1993). *Decision analysis and behavioral research.* Cambridge, MA: Cambridge University Press.

Vreeker, R., Nijkamp, P., & Ter Welle, C. (2002). A multicriteria decision support methodology for evaluating airport expansion plans. *Transportation Research Part D: Transport and Environment, 7*(1), 27–47.

Weng, S. Q., Huang, G. H., & Li, Y. P. (2010). An integrated scenario-based multi-criteria decision support system for water resources management and planning—A case study in the Haihe River Basin. *Expert Systems with Applications, 37*(12), 8242–8254.

Wierzbicki, A. P. (1992). Multi-objective modeling and simulation for decision support. Working paper of the International Institute for Applied Systems Analysis, WP-92-80. Laxemburg, Austria.

Xidonas, P., Mavrotas, G., Zopounidis, C., & Psarras, J. (2011). IPSSIS: An integrated multicriteria decision support system for equity portfolio construction and selection. *European Journal of Operational Research, 210*(2), 398–409.

Yager, R. R. (1993). Non-numeric multi-criteria multi-person decision making. *Group Decision Negotiation, 2,* 81–93.

Zeleny, M. (1974). *Linear multiobjective programming.* New York: Springer.

Zeleny, M. (1982). *Multiple criteria decision making.* New York: Mc Graw-Hill.

Zeng, Y., & Trauth, K. M. (2005). Internet-based fuzzy multicriteria decision support system for planning integrated solid waste management. *Journal of Environmental Informatics, 6*(1), 1–15.

Zhang, G., Ma, J., & Lu, J. (2009). Emergency management evaluation by a fuzzy multi-criteria group decision support system. *Stochastic Environmental Research and Risk Assessment, 23*(4), 517–527.

Zionts, S., & Wallenious, J. (1976). An interactive programming method for solving the multiple criteria problem. *Management Science, 22,* 652–663.

Zionts, S., & Wallenious, J. (1983). An interactive multiple objective linear programming method for a class of underlying non-linear utility functions. *Management Science, 29,* 512–529.

Zopounidis, C. (2001). *Analysis of financing decisions using multiple criteria* (pp. 67–85). Thessaloniki: Anikoula Publications.

Zopounidis, C., & Doumpos, M. (2002a). Multi-criteria decision aid in financial decision making: Methodologies and literature review. *Journal of Multi-Criteria Decision Analysis, 11*(4–5), 167–186.

Zopounidis, C., & Doumpos, M. (2002b). Multicriteria classification and sorting methods: A literature review. *European Journal of Operational Research, 138*(2), 229–246.

Zopounidis, C., Lemonakis, C., Andreopoulou, Z., & Koliouska, C. (2014). Agrotourism industry development through internet technologies: A multicriteria approach. *Journal of Euromarketing, 23*(4), 45–67.

Chapter 2
Methodologies

In this section, we will develop three methodologies: the PROMETHEE II ranking method, the Hierarchical Cluster Analysis and the K-means analysis.

2.1 Ranking Methods

Multicriteria analysis, often called multiple criteria decision making (MCDM) by the American School and multicriteria decision aid (MCDA) by the European School, is a set of methods which allow the aggregation of several evaluation criteria in order to choose, rank, sort or describe a set of alternatives (Zopounidis 1999). The one basic conviction underlying every MCDA approach is that the explicit introduction of several criteria, each representing a particular dimension of the problem to be taken into account, is a better path for robust decision-making when facing multidimensional and ill-defined problems, than optimizing a single-dimensional objective function (such as cost-benefit analysis) (Costa et al. 1997). In contrast to the more classical operations research approaches, the multicriteria decision aid framework facilitates learning about the problem and the alternative courses of action, by enabling people to think about their values and preferences from several points of view (Costa et al. 1997).

Interest in MCDA increased as the sphere of application of quantitative management science moved from operational decision making situations, for which a more-or-less well-defined single objective function could be identified with little controversy (e.g. maximize profit), to more complex levels of managerial planning and decision making, which are naturally multidimensional problems (Costa et al. 1997). Multicriteria analysis now supports the structuring of a decision problem, the exploration of the concerns of decision actors, the evaluation of alternatives under different perspectives and the analysis of their robustness against uncertainty (Beinat and Nijkamp 1998).

MCDA can deal with mixed sets of data, quantitative and qualitative, including expert opinions (Mendoza and Martins 2006). The capability to accommodate these gaps in information and knowledge through qualitative data, expert opinions, or experiential knowledge is a distinct advantage (Mendoza and Martins 2006). This participatory environment accommodates the involvement and participation of multiple experts and stakeholders (Mendoza and Prabhu 2003).

As various alternatives are available, selecting the best among them which satisfies the manufacturer's requirement becomes more complex and time consuming. To choose an appropriate material with several criterion is a multi-objective task and it is a multi criteria decision making (MCDM) problem (Anojkumar et al. 2016). Many multivariate techniques are extensions of univariate analysis (analysis of single-variable distributions) and bivariate analysis (cross-classification, correlation, analysis of variance and simple regression used to analyze two variables) (Hair 2010). For example, simple regression (with one predictor variable) is extended in the multivariate case to include several predictor variables (Hair 2010). Likewise, the single dependent variable found in analysis of variance is extended to include multiple dependent variables in multivariate analysis of variance (Hair 2010). Some multivariate techniques (e.g. multi le regression and multivariate analysis of variance) provide a means of performing in a single analysis what once took multiple univariate analyses to accomplish (Hair 2010). Other multivariate techniques, however, are uniquely designed to deal with multivariate issues, such as factor analysis, which identifies the structure underlying a set of variables, or discriminant analysis, which differentiates among groups based on a set of variables (Hair 2010).

The PROMETHEE (Preference Ranking Organization Method for Enrichment Evaluation) belongs to the class of MCDA instruments. Several MCDA techniques have been developed over the years that deal with the ranking of numerous alternatives based on a variety of criteria. MCDA can be efficiently utilized in order to handle both qualitative and quantitative criteria (Stiakakis and Sifaleras 2013). In other words, the MCDA allows for the selection of the best from the analyzed alternatives. Their development was actually the result of the practitioner's motivation to provide academics and researchers with improved decision making processes suitable for real-life multiple criteria decision situations by taking advantage of the recent evolutions in computer technology and the mathematical techniques involved (Wiecek et al. 2008). An important objective of outranking methods is that the decision maker should be able to incorporate any argument into the process to achieve better awareness of the decision problem (Lerche and Geldermann 2015). In this book, following Kosmidou and Zopounidis (2008a, b), we will present one of the most recent MCDA techniques, the PROMETHEE II method.

2.1.1 PROMETHEE II Method

PROMETHEE II method is an outranking multi-criteria decision aid approach developed and presented for the first time by Brans (1982) at the University Laval, Quebec, Canada, during an organized conference on multi-criteria decision aid instruments by Nadeau and Landry. This method has attracted the increased attention of the researchers for practically complex problems and the growing records of conference presentations and academic papers can easily illustrate this. As the time passed, a number of extensions have been suggested with the aim of assisting researchers in dealing with more complex problems. Indeed, PROMETHEE methodology has effectively been applied in a variety of areas such as Banking, Business and Financial Management, Chemistry, Energy resources, Health, Investments, Industrial Location, and other fields. As Brans and Mareschal (2005) have pointed out, the above technique owes its success mainly to its particular friendliness of use and to its' mathematical properties.

Addressing a classification problem requires the development of a classification model that aggregates the characteristics of the alternatives to provide recommendations on the assignment of the alternatives to the predefined classes. The significance of classification problems has motivated the development of a plethora of techniques for constructing classification models. Statistical techniques have been dominating the field for many years, but during the last two decades other approaches have become popular mainly from the field of machine learning (Andreopoulou et al. 2014a).

Also, the contributions of MCDA are mainly focused on the study of multicriteria classification problems (MCPs). MCPs can be distinguished from traditional classification problems studied within the statistical and machine learning framework in two aspects (Zopounidis and Doumpos 2002). The first aspect involves the nature of the characteristics describing the alternatives, which are assumed to have the form of decision criteria providing not only a description of the alternatives but also some additional preferential information. The second aspect involves the nature of the predefined classification which is defined in ordinal rather than nominal terms. Classification models developed through statistical and machine learning techniques often fail to address this issues focusing solely on the accuracy of the results obtained from the model.

Within the MCDA several criteria aggregation forms have been proposed for developing decision models. These include relational forms, value functions, and rule-based models.

The PROMETHEE II method is part of the outranking relations theory (Brans and Vinke 1985; Brans et al. 1986; Siskos and Zopounidis 1987; Brans et al. 1987; Brans et al. 1998; Zopounidis 2001). Relational models are based on the construction of an outranking relation that is used to compare the alternatives with some reference profiles characterizing each class. The reference profiles are either typical examples (alternatives) of each class or examples that define the upper/lower bounds of the classes. Some typical examples of this approach include methods such as ELECTRE TRI (Roy and Bouyssou 1993), PROAFTN (Belacel 2000), PAIRCLAS (Doumpos and Zopounidis 2004), and PROMETHEE TRI (Figueira et al. 2004). The main advantage of this approach is that it enables the decision maker (DM) to take into account the non-compensatory character of the decision process and to identify alternatives with special characteristics through the incorporation of the incomparability relation in the analysis. On other hand, the construction of the outranking relation requires the specification of a considerable amount of information which is not always easy to obtain (Andreopoulou et al. 2014a; Zopounidis et al. 2014). Brans (2015) presents a brief overview of OR from 1937 till now, the role of the outranking procedures in the field of MCDA and the PROMETHEE methodology including its extensions being of course stressed.

Value functions have also been quite popular as a criteria aggregation model in classification problems. This approach provides a straightforward methodology to perform the classification of the alternatives. Each alternative is evaluated according to the constructed value function and its global evaluation is compared to some value cut-off points in order to perform the assignment to one of the predefined classes. Due to their simplicity linear or additive value functions are usually considered (Jacquet-Lagreze 1995; Zopounidis and Doumpos 1999, 2000; Lemonakis and Strikos 2012). These provide a simple evaluation mechanism which is generally easy to understand and implement. However, there has been criticism on the assumptions underlying the use of such simple models and their ability to capture the interactions between the criteria.

The PROMETHEE methodology gives the researcher the ability to solve a decision problem where a finite set of comparable alternatives is to be evaluated according to several and often opposing criteria. The implementation of the PROMETHEE method involves the construction of an evaluation table (Table 2.1), in which the alternatives are estimated on the preferred criteria and ranked from the best to the worst. The PROMETHEE methods are considered to provide solutions for multicriteria problems of the form (2.1) and their associated evaluation table.

$$\max\{\ g1(a), g2(a), g3(a), \ldots, gj(a), \ldots, gk(a) | a \in A\ \} \qquad (2.1)$$

where:

A is a finite set of possible alternatives $\{\alpha 1, \alpha 2, \ldots, \alpha i, \ldots, \alpha n\}$ ($\alpha_1, \alpha_2, \ldots, \alpha_i, \ldots \alpha_n$) & $\{g1(*), g2(*), \ldots, gj(*), \ldots, gk(*)\}$ is a set of evaluation criteria.

Additional requirements for the application of PROMETHEE are the consideration of the relative significance of the selected criteria (i.e., the weights) and the information on the individually defined preference function of the decision-maker, regarding the comparison of the alternatives in terms of each single criterion.

The weights are typically arbitrary positive numbers, determined independently from the measurement units of the criteria. These numbers actually represent the relative significance of each criterion. The higher becomes the value of the weight, the higher the significance of the relevant criterion and conversely. According to Macharis et al. (2004), the selection of the weights is of high importance in the case

2.1 Ranking Methods

Table 2.1 Evaluation Table

a	g1(*)	g2(*)	...	gj(*)	...	gk(*)
α1	g1(α1)	g2(α1)	...	gj(a1)	...	gk(a1)
α2	g1(α2)	g2(α2)	...	gj(a2)	...	gk(a2)
.	.	.	.	.	.	.
.	.	.	.	.	.	.
.	.	.	.	.	.	.
αi	g1(αi)	g2(αi)	...	gj(ai)	...	gk(ai)
.	.	.	.	.	.	.
.	.	.	.	.	.	.
.	.	.	.	.	.	.
αn	g1(αn)	g2(αn)	...	gj(an)	...	gk(an)

Source Brans and Mareschal (2005)

of multicriteria decision analysis, since it reflects the decision-makers' insights and priorities.

The preference structure of PROMETHEE is based on pair wise comparisons. This means that a separate preference function for each criterion must be defined for all pairs of alternatives, reflecting the degree of preference for an alternative a over b. Vincke and Brans (1985) suggested six specific types of preference functions, provided in the appendix section, from which the researcher can easily define its preference structure. No matter, which is the preference function, the decision maker has to define the values of q, p and σ parameters. In contrast to q which is an indifference threshold that corresponds to the largest deviation, p is a strict preference threshold with the smallest deviation, capable of generating a full preference sufficiently for the decision maker. As far as the σ parameter is concerned it represents an intermediate value between q and p.

According to Brans et al. (1986), this preference degree for all couples of actions, can be represented by the preferred index of the following form:

$$\Pi(\alpha, \beta) = \frac{\sum_{j=1}^{n} w_j P_j(a,b)}{\sum_{j=1}^{n} w_j}$$

where:
wj is the weight for each criterion
Pj (a, b) expresses the degree at which bank a is preferred to bank b, when all the criteria are considered at once. Its value varies between 0 and 1.

A value equal to unity for the index will imply a strong preference of a bank a over b, while a zero value will imply a weak preference respectively. From the preference functions described above, this study utilized the Gaussian form for all the selected criteria. This function requires only for the parameter σ to be specified and at the same time, due to the lack of discontinuities, it gives robust and stable results.

As for the ranking of alternative actions, two flows should be defined, the leaving and the entering flow, briefly described below:

$$\Phi^+(\alpha) = \sum_{b \in X} \pi(a, b)$$

$$\Phi^-(\alpha) = \sum_{b \in X} \pi(b, a)$$

where:
X is the total of alternative solutions

The leaving flow $\phi+(a)$ expresses how an alternative a dominates all the other alternatives of X (the outranking character of a). On the other hand, the entering flow $\phi-(a)$ measures how an alternative a is surpassed by all the other alternatives of X (the outranked character of a). According to PROMETHEE I partial ranking an action a is favoured over an action b, (aPb) if the leaving and entering flows of action a are greater and smaller respectively than those of action b:

aPb if : $\phi+(a) > \phi+(b)$ and $\phi-(a) < \phi-(b)$ or
$\phi+(a) > \phi+(b)$ and $\phi-(a) = \phi-(b)$ or
$\phi+(a) = \phi+(b)$ and $\phi-(a) < \phi-(b)$.

In the case that the leaving and entering flows of two actions a and b are the same, the indifference situation can be written with the following expression (aIb):

aIb if : $\phi+(a) = \phi+(b)$ and $\phi-(a) = \phi-(b)$

There is also the possibility for two alternative actions to be incomparable, (aRb), if the entering flow of action a is worse than the corresponding flow of action b, while the opposite is implied by the leaving flow:

aRb if : $\phi+(a) > \phi+(b)$ and $\phi-(a) > \phi-(b)$ or $\phi+(a) < \phi+(b)$ and $\phi-(a) < \phi-(b)$

In this paper we utilized only the PROMETHEE II method which provides a complete ranking of the comparable alternatives from the best to the worst. The net flow implied by $\Phi(a)$, which is the difference between the two flows, corresponds to a value function for which the higher the value the higher the attractiveness of alternative a. For each action a $\in$ X the net flow can be described as follows:

$$\Phi(a) = \Phi+(a) - \Phi-(a)$$

2.1 Ranking Methods

The outranking relations in PROMETHEE II method are such that:

$$\alpha P^{II} b \quad \text{if} \quad \Phi(a) > \Phi(b),$$
$$\alpha I^{II} b \quad \text{if} \quad \Phi(a) = \Phi(b)$$

When, $\alpha P^{II} b$, alternative α is preferred over b.

Also, when $\alpha I^{II} b$, the decision maker is indifferent between alternatives α and b.

The net flow is the number that is used for the comparison between the cases in order to obtain the ranking. Each case with a higher net flow is considered superior in ranking.

In fact, there exist two types of the PROMETHEE methodology, the PROMETHEE I that ranks partially and also, the PROMETHEE II, which performs a full and complete ranking, based on all of the input data. In contrast to PROMETHEE I, incomparabilities are now absent between the alternatives. As a result, the alternative with the higher net flow is identified as the one optimizing all the criteria.

In general, the PROMETHEE methods include the three following steps (Brans and Mareschal 1994):

1. Enrichment of the preference structure. The notion of generalized criteria is introduced in order to take into account the amplitudes of the deviations between the evaluations. This step is crucial. Yet it can easily be understood by the decision maker because all the additional parameters to be defined have an economical significance. Moreover, the scaling effects are entirely handled in this first step.
2. Enrichment of the dominance relation. A valued outranking relation is built taking into account all the criteria. For each pair of alternatives, the overall degree of preference of one alternative over the other is obtained.
3. Exploitation for decision aid. PROMETHEE I provides a partial ranking of A, including possible incomparabilities. PROMETHEE II provides a complete ranking of A. It can look more efficient but in fact the information used is more disputable.

Multiattribute utility theory is not the only framework about multicriteria decision problems. In this approach, the cardinal approach, we aggregate numbers (the monodimensional utilities) representing an absolute evaluation of a given alternative with respect to a given criterion (Grabish 1995). As for the relational approach, the alternatives are compared two by two and the degree of preference of one alternative over the other, with respect to a criterion (relative evaluation) is expressed with a number. All these preference relations are then aggregated to take into account all the criteria (Grabish 1995). This approach has been developed essentially by Roy (1968, 1972) (ELECTRE methods) with ordinary crisp relations and then by Fodor and Roubens (1992) with fuzzy preference relations.

In Chap. 3, three examples will be presented using PROMETHEE II method. The first case study deals with skiing centers, the second case study is about 20 agrotourism enterprises and the third one regards 20 aquaculture units. These entities are ranked according to certain qualitative characteristics.

2.2 Cluster Generation Methods

Cluster analysis describes a set of multivariate methods and techniques that seek to classify data, often into groups, types, profiles, and so on (Leonard and Droege, 2008). For example, cluster analysis can be used to develop taxonomies or typological frameworks, to explore data to unravel complex underlying patterns, and may also be understood as a type of data reduction procedure (Leonard and Droege). Because of its utility, and due to advances in software technology, studies using cluster analysis have been surging in number since its introduction in the early 1960s (Leonard and Droege). Aldenderfer and Blashfield (1984) suggest that the popularity of cluster analysis can be attributed to its applicability to all scientific disciplines that rely on classification systems to guide research and understanding (Leonard and Droege). Cluster analysis is relevant to research in biology, medicine, education, archaeology, psychology, and other sciences because all such scientific disciplines rely in some way on classification (Leonard and Droege).

Classification refers to the assignment of a finite set of alternatives into groups. The alternatives belonging into different groups have different characteristics, without being possible to establish any kind of preference relation between them (i.e. the groups provide a description of the alternatives without any further information) (Doumpos and Zopounidis 2002).

Cluster analysis strategy is to detect structures, non visible and will always form clusters with the objects not considering the true existence of any structure in the data (Arabatzis and Kokkinakis 2005). The cluster solution is completely dependent among the variables used as the basis for the similarity measure and the method. When a cluster solution is identified the researcher should furthermore examine the fundamental structure represented in the defined clusters. Moreover, in the case that a single member cluster appears, the researcher must decide if it represents a valid structural component in the sample or if it should be deleted as unrepresentative (Hair et al. 1998). The formation of clusters/groups is based on ad hoc simple calculative procedures (Kinnear and Taylor 1996), although with substantial mathematical identities, they are no more than clever algorithms, and their result has to be translated with practical rules, mainly objective.

Although cluster analysis is conceptually simple to understand, some features of cluster analysis methods can be confusing (Leonard and Droege). This is largely due to the diversity of applications of cluster analysis across disciplines (Kogan 2007); many of the evolutionary stages of cluster analysis have occurred in silos and are somewhat discipline specific, without much cohesion or translation of developments and findings across disciplines (Leonard and Droege). There is no

2.2 Cluster Generation Methods

detailed, uniform protocol (i.e., step-by-step progression) in conducting cluster analysis, which makes it different from multivariate procedures such as multivariate analysis of variance (MANOVA) and factor analysis (Leonard and Droege). The researcher using cluster analysis for the first time will likely be confronted with a set of choices and options that will need to be customized to her or his own unique research problem (Leonard and Droege). This is not to suggest that cluster analysis is a haphazard collection of procedures (Leonard and Droege). In fact, several authors have proposed guidelines for conducting CA in a step-by-step manner (Leonard and Droege).

According to Milligan (1996, 2007), there are seven steps in conducting cluster analysis:

1. Composition of the dataset
2. Selection of variables
3. Decisions about standardizing variables
4. Selecting a measure of association (or similarity measure)
5. Selecting a clustering method
6. Determining the number of clusters
7. Interpretation, testing and replication

Within each step, there are multiple options for implementation (Leonard and Droege). Unless one is an expert in cluster analysis or has a thorough statistical background, choosing from among the possible options can be challenging (Leonard and Droege). At this point in the evolution of cluster analysis, it is rare for a researcher to find a straightforward, direct path through the series of steps involved in completing cluster analysis (Leonard and Droege). This is not intended to discourage researchers from trying cluster analysis, but it may be advisable to anticipate these challenges and not be alarmed if and when they emerge (Leonard and Droege). It is also important to clearly state that it is unlikely that this article will provide a pharmacy researcher with all the information he or she would need to successfully complete cluster analysis; chances are that one would need to consult additional sources directly, although the reference list contains many useful sources (Leonard and Droege).

Cluster analysis can be performed either with the technique of hierarchical cluster analysis or with the technique of k-means cluster analysis (Everrit 1993). Furthermore, both hierarchical clustering and k-means generate "hard" solutions that define partitions of the data (Ahlquist and Breunig 2012). There is no foundation in statistical theory on which to prefer a particular clustering solution over another and no possibility of evaluating the uncertainty around a particular observation's assignment to a given cluster (Ahlquist and Breunig 2012). The choice of both the number of clusters to focus on and the substantive interpretations assigned to them is solely the responsibility of the analyst (Ahlquist and Breunig 2012).

Initially, cluster randomized designs were widely regarded as lacking statistical precision (Raudenbush 1997). Cluster analysis is a descriptive method, used mostly as an exploratory implement that without any statistical basis at which to draw statistical assumptions from a sample to a population (Hair et al. 1998). In most real life clustering situations, an applied researcher is faced with the dilemma of selecting the number of clusters or partitions in the final solution (Everitt 1979; Sneath and Sokal 1973). Virtually all clustering procedures provide little if any information as to the number of clusters present in the data (Milligan and Cooper 1985). A theoretical consideration which underlies many of the hierarchical methods is that some of these algorithms invoke the ultrametric inequality in the solution process (Milligan 1980). Pvclust is an add-on package for a statistical software R to assess the uncertainty in hierarchical cluster analysis (Suzuki and Shimodaira 2006). Pvclust can be used easily for general statistical problems, such as DNA microarray analysis, to perform the bootstrap analysis of clustering, which has been popular in phylogenetic analysis (Suzuki and Shimodaira 2006). Most (although by no means all) investigators are now warier of the whole area, having become aware of the varied and difficult problems facing the cluster analysis user in practice (Everitt 1979). Various qualitative guidelines have been proposed for deciding at what point in the clustering process clusters become non-significant (e.g., Thorndike 1953; Marriot 1971; Mojena 1977; Struass 1982). Everitt (1980), noting that the determination of the number of significant groups in a cluster analysis is a "formidable problem," has identified three principal difficulties encountered in deriving adequate significance tests: (1) specification of a suitable null hypothesis; (2) determination of the sampling distribution of the distance or similarity measure used; and (3) development of a flexible test procedure.

2.2.1 *Hierarchical Cluster Analysis*

Hierarchical cluster analysis produces a unique set of nested categories or clusters by sequentially pairing variables, clusters, or variables and clusters (Bridges 1966). At each step, beginning with the correlation matrix, all clusters and unclustered variables are tried in all possible pairs, and that pair producing the highest average intercorrelation within the trial cluster is chosen as the new cluster (Bridges 1966). In contrast to other types of cluster analysis in which a single set of mutually exclusive and exhaustive clusters is formed, this technique proceeds sequentially from tighter, less inclusive clusters through larger more inclusive clusters and is continued until all variables are clustered in a single group (Bridges 1966). A graph, constructed like the taxonomic dendrogram of the biological systematist, shows the class-inclusive relations between clusters and the value of the clustering criterion associated with each (Bridges 1966).

An important characteristic of hierarchical procedures is that the results at an earlier stage are always nested within the results at a later stage, creating a similarity to a tree (Arabatzis and Kokkinakis 2005). However, divisions or fusions of clusters once made are irrevocable, so that when an agglomerative algorithm has joined two objects they cannot subsequently be separated. The disadvantage of this technique lies in the fact that an exchange of elements between the groups is impossible when the "tree structure" is building up (Gerstengarbe et al. 1999).

Hierarchical cluster analysis is primarily an exploratory rather than confirmatory or inferential activity. In fact, Kaufman and Rousseeuw (2005) suggest that "it is permissible to try several algorithms on the same data because cluster analysis is mostly used as a descriptive or exploratory tool....we just want to see what the data are trying to tell us." There are many attributes on which to measure similarity and difference across objects and, given some set of attributes, numerous algorithms for identifying clusters (Ahlquist and Breunig 2012).

Hierarchical analysis uses intuitively plausible procedures based on various distance metrics to either merge or partition observations into clusters (Ahlquist and Breunig 2012).

Hierarchical procedure involves the construction of a hierarchy of a tree like structure that helps to indicate the number of clusters within the cases (Andreopoulou et al. 2014b). In order to determine the final number of the clusters (stopping rule) large increases in the average within-cluster distance are identified. The prior cluster solution is then selected on the logic that its combination caused a substantial decrease in similarity (Hair et al. 1998).

Hierarchical cluster analysis is a method for finding the underlying structure of objects through an iterative process that associates (agglomerative methods) or dissociates (divisive methods) object by object, and that is halted when all objects have been processed (Steinbach et al. 2003; Almeida et al. 2007). The agglomerative procedure starts with each object in a separate cluster and then combines the clusters sequentially, reducing the number of clusters at each step until all objects belong to only one cluster (Almeida et al. 2007).

It begins with each object on its own and proceeding to combine them into clusters that maximize within-cluster similarity and between-cluster difference, as determined by a distance metric. Several different metrics can be employed and the literature provides little theoretical guidance about their appropriateness, though Milligan (1980, 1981) surveys Monte Carlo experiments, concluding that Ward's linkage is a useful distance metric.

Among the five most popular agglomerative algorithms used to develop clusters is the Ward's method. In Ward's method is a clustering procedure seeking to form the partitions in a manner that minimizes the loss associated with each grouping and to quantify that loss in terms of an error sum-of-squares criterion (ESS) (Everrit 1993). Ward's method is designed to form clusters that have minimum within-cluster sum-of-squares (Waller et al. 1998). The complete linkage method forms clusters in which each entity is more similar to other members of its cluster than to all members of other clusters (Waller et al. 1998). The average linkage method forms clusters that each entity has a greater mean similarity with all members of its cluster than to all members of other clusters (Waller et al. 1998).

The divisive methods start with all of the objects in one cluster, and then proceed to their partition into smaller clusters until there is one object per cluster (Downs and Barnard 2002; Bratchell 1989; Almeida et al. 2007). This means that for N objects, the process involves $N - 1$ clustering steps.

In hierarchical cluster analysis there are two important choices when defining a method: the type of similarity measure between objects and/or groups, and the linkage technique (Bratchell 1989). The first task is to determine a numerical value for the similarity between objects, constructing a similarity matrix (Almeida et al. 2007). The most popular ways to determine the similarity between objects use the Euclidean distance and the correlation coefficient, but there are many alternatives for similarity indicators (Kellner et al. 2004; Brereton 2004).

The next step is to group or ungroup the objects. The most common approach is an agglomerative technique, whereby single objects are gradually connected to each other in groups. The first connection corresponds necessarily to the most similar pair of objects. Once the first group is formed, it is necessary to define the similarity between the new group and the remaining objects (Brereton 2004). This step requires a new choice among a variety of available techniques. Some of the most used linkage algorithms are complete-linkage (or furthest-neighbor), single linkage (or nearest-neighbor), average-linkage (between groups and within groups), centroid method and Ward's-linkage (Downs and Barnard 2002; Kellner et al. 2004; Smolinski et al. 2002). In single linkage, when a new group is formed, the corresponding distance to any other group is the minimal Euclidean distance of all possible distances between each object of the former group and each object of the latter (Almeida et al. 2007).

Once the similarity measure and the linkage method are defined, the agglomeration of objects and groups in each step of the process follows the order of larger similarity (Brereton 2004). The structure obtained by hierarchical clustering is often presented in the form of a dendrogram where each linkage step in the clustering process is represented by a connection line (Downs and Barnard 2002; Smolinski et al. 2002).

It is widely accepted that the average-linkage, centroid and Ward's methods are sensitive to the shape and size of clusters. Thus, they can easily fail when clusters have complicated forms departing from the hyperspherical shape (Downs and Barnard 2002). Complete-linkage is not strongly affected by outliers, but can break large clusters, and has trouble with convex shapes (Steinbach et al. 2003). The single linkage methodology, on the other hand, displays total insensibility to shape and size of clusters (Downs and Barnard 2002). However, there are also shortcomings associated with single linkage, which is the sensitivity to the presence of outliers and the difficulty in dealing with severe differences in the density of clusters (Almeida et al. 2007).

Hierarchical cluster analysis can further be subdivided into 2 methods: agglomerative and divisive (Leonard and Droege). Agglomerative methods use algorithms to join cases based on their similarity, and divisive methods use algorithms that separate cases based on their differences (Leonard and Droege). With either method, hierarchical cluster analysis is sequential (Leonard and Droege). Essentially, the hierarchical algorithm sorts through the dataset in a serial manner, seeking to divide (divisive methods) or fuse (agglomerative methods) 2 cases at a time (Leonard and Droege). This also means that when cases are joined or separated, hierarchical cluster analysis methods do not return to those 2 cases again

(Leonard and Droege). This is where the term "hierarchical" applies: Once a match or separation is achieved, it remains in place for the remainder of the analysis (Leonard and Droege). This process continues until all cases are assigned to a cluster. hierarchical cluster analysis involves many more computations than nonhierarchical methods, and this may be an issue with very large datasets (Leonard and Droege). Advances in computer technology may eventually obviate this consideration, but for now, extremely large datasets may be inappropriate for hierarchical cluster analysis methods (Aldenderfer and Blashfield 1984).

The application of different methods, which may involve different similarity measures, different linkage techniques, etc., leads to dendrograms with different structures (Almeida et al. 2007). Apparently, a good approach would be to use different methods of cluster analysis and compare the results, but due to an excessive wealth of options it is frequently more convenient to use well founded a priori choices (Almeida et al. 2007).

In Chap. 3, two examples will be presented using hierarchical cluster analysis. The first case study deals with 20 wood enterprises and the second one is about 20 enterprises which promote the Renewable Energy Sources. These entities are classified according to certain qualitative and quantitative characteristics.

2.2.2 K-Means Analysis

K-means cluster analysis is an example of a nonhierarchical cluster analysis method, where k is equal to the number of clusters the researcher wishes to impose upon the data (Leonard and Droege). Because k is determined a priori, k-means cluster analysis does not "suffer" from the issue of imposing a single partition only once and never visiting the partition again (Leonard and Droege).

Within traditional cluster analysis, relocation methods such as k-means require that the analyst posit the number of clusters in the data in advance and then proceed to iteratively move observations among clusters until an optimal allocation can be identified (Ahlquist and Breunig 2012). For example, k-means iteratively moves observations from one cluster to another to minimize the total squared distance from k "centroids" or "prototypes" (Ahlquist and Breunig 2012).

A generic outline that describes all k-means algorithms is presented below (Blashfield and Aldenderfer 1988; Darken and Moody 1990; Waller et al. 1998):

1. Initialize the seed values for a prespecified number (k) of clusters. These seed values represent the cluster centroids. If computationally feasible, cluster centroids from a prior hierarchical cluster analysis (e.g. Ward's method or group average) can be used as seeds. This method of initializing the k-means seed values has been recommended by Milligan, who—when summarizing the findings of a previous Monte Carlo study—concluded that "the k-means algorithms do no seem to be very desirable if random starting seeds must be used"

2. Allocate each data point in the sample to the cluster with the nearest centroid. Proximity is defined using Euclidean distances.
3. If a cluster increased in size during the last data pass, that is, if new data points were allocated to the cluster, then recomputed the cluster centroid.
4. Alternate steps 2 and 3 until no data points change clusters.

The aim of the k-means algorithm is to generate clusters with minimum within-cluster sum of squares (Hartigan 1975). Unfortunately, there is no guarantee that this aim will be realized when the number of input vectors is moderate or large or when the number of clusters is larger than two (Waller et al. 1998). With more than two clusters there is an exponentially increasing number of initial data partitions from which to start the algorithm (Waller et al. 1998). The Hartigan and Wong (1979) algorithm that was used in the present study finds local optima such that, given the initial cluster seed values, the obtained clusters have minimum sums of squares. Alternative seeds may produce tighter clusters and thus the quality of the solution is sensitive to the initial cluster centroids (Waller et al. 1998).

The K-means algorithm requires three user-specified parameters: number of clusters K, cluster initialization, and distance metric (Jain 2010). The most critical choice is K (Jain 2010). While no perfect mathematical criterion exists, a number of heuristics are available for choosing K (Jain 2010). Typically, K-means is run independently for different values of K and the partition that appears the most meaningful to the domain expert is selected (Jain 2010). Different initializations can lead to different final clustering because K-means only converges to local minima (Jain 2010). One way to overcome the local minima is to run the K-means algorithm, for a given K, with multiple different initial partitions and choose the partition with the smallest squared error (Jain 2010).

K-means is typically used with the Euclidean metric for computing the distance between points and cluster centers (Jain 2010). As a result, K-means finds spherical or ball-shaped clusters in data (Jain 2010). K-means with Mahalanobis distance metric has been used to detect hyper-ellipsoidal clusters (Mao and Jain 1996), but this comes at the expense of higher computational cost (Jain 2010). A variant of K-means using the Itakura–Saito distance has been used for vector quantization in speech processing (Linde et al. 1980) and K-means with L1 distance was proposed in (Kashima et al. 2008). Banerjee et al. (2004) exploits the family of Bregman distances for K-means.

K-Means is a standard technique for clustering, used in a wide array of applications and even as way to initialize the more expensive EM clustering algorithm (Bishop 1995; Cheeseman and Stutz 1996; Meila and Heckerman 1998). Furthermore, regardless of which clustering algorithm is being used, K-Means is employed internally by our initialization refinement method (Bradley and Fayyad 1998).

The k-means algorithm is defined over continuous data (Niknam and Amiri 2010). The k-means algorithm gave better results only when the initial partitions were close to the final solution (Niknam and Amiri 2010). In other words, the results of k-means highly depend on the initial state and reach to local optimal solution. In order to overcome this problem, a lot of studies have done in clustering

(Kao et al. 2008; Cao and Krzysztof 2008; Zalik 2008; Krishna 1999; Mualik and Bandyopadhyay 2000; Fathian and Amiri 2007; Laszlo and Mukherjee 2007; Shelokar et al. 2004; Ng and Wong 2002; Sung and Jin 2000; Niknam et al. 2008; Niknam et al. 2008; Niknam et al. 2008).

The k-means algorithm has the following important properties (Huang 1998):

5. It is efficient in processing large data sets.
6. It often terminates at a local optimum (MacQueen 1967; Selim and Ismail 1984).
7. It works only on numeric values.
8. The clusters have convex shapes (Anderberg 1973).

There exist a few variants of the k-means algorithm which differ in selection of the initial k means, dissimilarity calculations and strategies to calculate cluster means (Anderberg 1973; Bobrowski and Bezdek 1991). The sophisticated variants of the k-means algorithm include the well-known ISODATA algorithm (Ball and Hall 1967) and the fuzzy k-means algorithms (Ruspini 1969, 1973; Bezdek 1981). One difficulty in using the k-means algorithm is that the number of clusters has to be specified (Anderberg 1973; Milligan and Cooper 1985). Some variants like ISODATA include a procedure to search for the best k at the cost of some performance (Huang 1998).

In Chap. 3, three examples will be presented using k-means analysis. The first case study deals with 20 government agencies in National Parks, the second one is about the agricultural exploitations of 50 prefectures and the third one deals with 20 agrifood entities These entitles are classified according to certain quantitative and qualitative characteristics.

References

Ahlquist, J. S., & Breunig, C. (2012). Model-based clustering and typologies in the social sciences. *Political Analysis, 20*(1), 92–112.

Aldenderfer, M. S., & Blashfield, R. K. (1984). *Cluster analysis.* Newbury Park: Sage.

Almeida, J. A. S., Barbosa, L. M. S., Pais, A. A. C. C., & Formosinho, S. J. (2007). Improving hierarchical cluster analysis: A new method with outlier detection and automatic clustering. *Chemometrics and Intelligent Laboratory Systems, 87*(2), 208–217.

Anderberg, M.R. (1973). *Cluster analysis for applications.* New York: Academic Press.

Andreopoulou, Z., Tsekouropoulos, G., Koliouska, C., & Koutroumanidis, T. (2014a). Internet marketing for sustainable development and rural tourism. *International Journal of Business Information Systems, 16*(4), 446–461.

Andreopoulou, Z., Tsekouropoulos, G., Theodoridis, A., Samathrakis, V., & Batzios, C. (2014b). Consulting for sustainable development, information technologies adoption, marketing and entrepreneurship issues in livestock farms. *Procedia Economics and Finance, 9,* 302–309.

Anojkumar, L., Ilangkumaran, M., & Hassan, S. M. (2016). An integrated hybrid multi-criteria decision making technique for material selection in the sugar industry. *International Journal of Multicriteria Decision Making, 6*(3), 247–268.

Arabatzis, G. D., & Kokkinakis, A. K. (2005). Typology of the lagoons of Northern Greece according to their environmental characteristics and fisheries production. *Operational Research, 5*(1), 21–34.

Ball, G. H., & Hall, D. J. (1967). A clustering technique for summarizing multivariate data. *Behavioral Science, 12*, 153–155.
Banerjee, A., Merugu, S. Dhillon, I. & Ghosh, J. (2004). Clustering with bregman divergences. *Journal of machine learning research,* 234–245
Beinat, E., & Nijkamp, P. (1998). Multicriteria analysis for land-use management (Vol. 9). Dordrecht: Springer Science & Business Media.
Belacel, N. (2000). Multicriteria assignment method PROAFTN: Methodology and medical applications. *European Journal of Operational Research, 125,* 175–183.
Bezdek, J. C. (1981). *Pattern recognition with fuzzy objective function.* New York: Plenum Press.
Bishop, C. (1995). *Neural networks for pattern recognition.* New York: Oxford University Press.
Blashfield, R. K., & Aldenderfer, M. S. (1988). The methods and problems of cluster analysis. In *Handbook of multivariate experimental psychology* (pp. 447–473). New York: Springer.
Bobrowski, L., & Bezdek, J. C. (1991). c-Means clustering with the l1 and l∞ norms. *IEEE Transactions on Systems, Man and Cybernetics, 21*(3), 545–554.
Bradley, P. S., & Fayyad, U. M. (1998). Refining Initial Points for K-Means Clustering. In *ICML* (Vol. 98, pp. 91–99).
Brans, J.P. (1982). Lingenierie de la decision. Elaboration dinstruments daide a la decision. Methode PROMETHEE. In: Nadeau, R., & Landry, M. (eds.), Laide a la Decision: Nature, Instrument s et Perspectives Davenir (pp. 183–214). Quebec: Presses de Universite Laval.
Brans, J. P. (2015). The 'PROMETHEE' adventure. *International Journal of Multicriteria Decision Making, 5*(4), 297–308.
Brans, J. P., & Mareschal, B. (1994). The PROMCALC & GAIA decision support system for multicriteria decision aid. *Decision Support Systems, 12*(4–5), 297–310.
Brans, J. P. & Mareschal, B. (2005). PROMETHEE methods. In *Multiple criteria decision analysis: State of the art surveys* (pp. 163–186). New York: Springer.
Brans, N., & Vinke, Ph. (1985). A preference ranking organization method: The PROMETHEE method for multiple criteria decision making. *Management Science, 31*(6), 647–656.
Brans, J. P., Vinke, P., & Mareschal, B. (1986). How to select and how to rank projects: The PROMETHEE method. *European Journal of Operational Research, 24,* 228–238.
Brans, J. P., Mareschal, B., Margeta, J., & Mladineo, N. (1987). Multicriteria ranking of alternative locations for small scale hydro plants. *European Journal of Operational Research, 31,* 215–222.
Brans, J. P., Chevalier, A., Kunsch, P., Macharis, C., & Schwaninger, M. (1998). Combining multicriteria decision aid and system dynamics for the control of socio-economic processes. *European Journal of Operational Research, 109,* 428–441.
Bratchell, N. (1989). Cluster analysis. *Chemometrics and Intelligent Laboratory Systems,* 105–125.
Brereton, R. G. (2004). *Chemometrics, data analysis for the laboratory and chemical plant* (1st ed.). London: Wiley.
Bridges, C. C. (1966). Hierarchical cluster analysis. *Psychological Reports, 18*(3), 851–854.
Cao, D. N., & Krzysztof, J. C. (2008). GAKREM: a novel hybrid clustering algorithm. *Information Sciences, 178,* 4205–4227.
Cheeseman, P., & Stutz, J. (1996). *Bayesian classification (AutoClass): theory and results.* in [*FPSU96*] (pp. 153–180). Cambridge: MIT Press.
Costa, C. A. B. E., Stewart, T. J., & Vansnick, J. C. (1997). Multicriteria decision analysis: Some thoughts based on the tutorial and discussion sessions of the ESIGMA meetings. *European Journal of Operational Research, 99*(1), 28–37.
Darken, C., Moody, J. (1990). Fast adaptive k-means clustering: some empirical results. In *1990 IJCNN international joint conference on neural networks* (pp. 233–238). IEEE.
Doumpos, M., & Zopounidis, C. (2002). Multicriteria decision aid classification methods (vol. 73). New York: Springer Science & Business Media.
Doumpos, M., & Zopounidis, C. (2004). A multicriteria classification approach based on pairwise comparisons. *European Journal of Operational Research, 158,* 378–389.

References

Downs, G. M., & Barnard, J. M. (2002). Clustering methods and their uses in computational chemistry. *Reviews in Computational Chemistry, 18,* 1–40.

Everitt, B. S. (1979). Unresolved problems in cluster analysis. Biometrics, 169–181.

Everrit, B. (1993). *Cluster analysis* (3rd ed.). London-Sydney-Auckland: Arnold.

Fathian, M. & Amiri, B. (2007). A honey-bee mating approach on clustering. *The International Journal of AdvancedManufacturing Technology,* doi:10.1007/ s00170-007-1132-7

Figueira, J., De Smet, Y., & Brans, J. P. (2004). MCDA methods for sorting and clustering problems: Promethee TRI and Promethee CLUSTER. Université Libre de Bruxelles, Service de Mathématiques de la Gestion, Working Paper Feb 2004. (http://www.ulb.ac.be/polytech/smg/indexpublications.htm).

Fodor, J. C., & Roubens, M. (1992). Aggregation and scoring procedures in multicriteria decision making methods. In *IEEE international conference on fuzzy systems* (pp. 1261–1267). IEEE.

Gerstengarbe, F. W., Werner, P. C., & Fraedrich, K. (1999). Applying non-hierarchical cluster analysis algorithms to climate classification: some problems and their solution. *Theoretical and Applied Climatology, 64*(3–4), 143–150.

Grabisch, M. (1995). Fuzzy integral in multicriteria decision making. *Fuzzy Sets and Systems, 69* (3), 279–298.

Hair, J. (2010). Multivariate data analysis. Pearson College Division.

Hair, J., Anderson, R., Tatham, R., & Black, W. (1998). *Multivariate data analysis with readings* (5th ed.). Upper Saddle River: Prentice-Hall International, Inc.

Hartigan, J. A. (1975). *Clustering algorithms.* New York: Wiley.

Hartigan, J. A., & Wong, M. A. (1979). Algorithm AS 136: A k-means clustering algorithm. *Journal of the Royal Statistical Society: Series C (Applied Statistics), 28*(1), 100–108.

Huang, Z. (1998). Extensions to the k-means algorithm for clustering large data sets with categorical values. *Data Mining and Knowledge Discovery, 2*(3), 283–304.

Jacquet-Lagreze, E. (1995). An application of the UTA discriminant model for the evaluation of R&D projects. In P. M. Pardalos, Y. Siskos, & C. Zopounidis (Eds.), *Advances in multicriteria analysis* (pp. 203–211). Dordrecht: Kluwer Academic Publishers.

Jain, A. K. (2010). Data clustering: 50 years beyond K-means. *Pattern Recognition Letters, 31*(8), 651–666.

Kao, Y. T., Zahara, E., & Kao, I. W. (2008). A hybridized approach to data clustering. *Expert Systems with Applications, 34*(3), 1754–1762.

Kashima, H., Hu, J., Ray, B., & Singh, M. (2008). K-means clustering of proportional data using L1 distance. In: *Proceedings of international conference on pattern recognition* (pp. 1–4).

Kaufman, L. & Rousseauw, P. J. (2005). Finding groups in data: An introduction to cluster analysis. New York:Wiley-Qnterscience

Kellner, R., Mermet, J. M., Otto, M., Valcarcel, M., & Widmer, H. M. (2004). *Analytical chemistry: a modern approach to analytical science* (pp. 176–189, Chap. 8) (2nd ed.). Weinheim: Wiley.

Kinnear, T. & Taylor, J. (1996). *Marketing research. An applied approach* (5th ed.) New York: McGraw-Hill, Inc.

Kogan, J. (2007). *Introduction to clustering large and high- dimensional data.* Cambridge: Cambridge University Press.

Kosmidou, K., & Zopounidis, C. (2008a). Predicting US commercial bank failures via a multicriteria approach. *International Journal of Risk Assessment and Management, 9*(1–2), 26–43.

Kosmidou, K., & Zopounidis, C. (2008b). Measurement of bank performance in Greece. *South-Eastern Europe Journal of Economics, 1,* 79–95.

Krishna, K., & Murty, M. N. (1999). Genetic K-means algorithm. IEEE Transactions on Systems, Man, and Cybernetics, Part B (Cybernetics), *29*(3), 433–439.

Laszlo, M., & Mukherjee, S. (2007). A genetic algorithm that exchanges neighboring centers for k-means clustering. *Pattern Recognition Letters, 28*(16), 2359–2366.

Lemonakis, C., & Strikos, I. (2012). Measurement of commercial banks performance in EU countries: A multi-criteria approach. In *Financial services: efficiency and risk management* (Chap. 4), Nova Publisher.

Leonard, S. T., & Droege, M. (2008). The uses and benefits of cluster analysis in pharmacy research. *Research in Social and Administrative Pharmacy, 4*(1), 1–11.

Lerche, N., & Geldermann, J. (2015). Integration of prospect theory into PROMETHEE-a case study concerning sustainable bioenergy concepts. *International Journal of Multicriteria Decision Making, 5*(4), 309–333.

Linde, Y., Buzo, A., & Gray, R. (1980). An algorithm for vector quantizer design. *IEEE Transactions on Communications, 28*(1), 84–95.

Macharis, C., Springael, J., De Brucker, K., & Verbeke, A. (2004). PROMETHEE and AHP: The design of operational synergies in multicriteria analysis. Strengthening PROMETHEE with ideas of AHP. *European Journal of Operational Research, 153*, 307–317.

MacQueen, J. B. (1967). Some methods for classification and analysis of multivariate observations. In: *Proceedings of the 5th berkeley symposium on mathematical statistics and probability* (pp. 281–297).

Mao, J., & Jain, A. K. (1996). A self-organizing network for hyper-ellipsoidal clustering (HEC). *IEEE Transactiuon on Neural Networks, 7*(January), 16–29.

Marriot, F. H. C. (1971). Practical problems in a method of cluster analysis. *Biometrics, 27*, 501–514.

Meila, M., & Heckerman, D. (1998). An experimental comparison of several clustering methods. Microsoft Research Technical Report MSR-TR-98-06, Redmond, WA.

Mendoza, G. A., & Martins, H. (2006). Multi-criteria decision analysis in natural resource management: a critical review of methods and new modelling paradigms. *Forest Ecology and Management, 230*(1), 1–22.

Mendoza, G. A., & Prabhu, R. (2003). Qualitative multi-criteria approaches to assessing indicators of sustainable forest resource management. *Forest Ecology and Management, 174*, 329–343.

Milligan, G. W. (1980). An examination of the effect of six types of error perturbation on fifteen clustering algorithms. *Psychometrika, 45*, 325–342.

Milligan, G. W. (1981). A review of Monte Carlo tests of cluster analysis. *Multivariate Behavioral Research, 16*, 379–407.

Milligan, G. W. (1996). Clustering validation: results and implications for applied analyses. In P. Arabie, L. J. Hubert, & G. DeSoete (Eds.), *Clustering and classification* (pp. 345–379). River Edge: World Scientific Press.

Milligan, G. W. (2007). Cluster analysis. In: S. Kotz, B. R Campbell, N. Balakrishnan, B. Vidokovic (eds.), *Encyclopedia of statistical sciences* (2nd ed.) http://www.mrw.interscience. wiley.com/emrw/9780471667193/home. Accessed Mar 29, 2007.

Milligan, G. W., & Cooper, M. C. (1985). An examination of procedures for determining the number of clusters in a data set. *Psychometrika, 50*(2), 159–179.

Mojena, R. (1977). Hierarchical grouping methods and stop- ping rules: an evaluation. *Computer Journal, 20*, 359–363.

Mualik, U., & Bandyopadhyay, S. (2000). Genetic algorithm-based clustering technique. *Pattern Recognition, 33*, 1455–1465.

Ng, M. K., & Wong, J. C. (2002). Clustering categorical data sets using tabu search techniques. *Pattern Recognition, 35*(12), 2783–2790.

Niknam, T., & Amiri, B. (2010). An efficient hybrid approach based on PSO, ACO and k-means for cluster analysis. *Applied Soft Computing, 10*(1), 183–197.

Niknam, T., Olamaie, J., & Amiri, B. (2008a). A hybrid evolutionary algorithm based on ACO and SA for cluster analysis. *Journal of Applied Science, 8*(15), 2695–2702.

Niknam, T., Firouzi, B. B., & Nayeripour, M. (2008). An efficient hybrid evolutionary algorithm for cluster analysis. In 2008 World Applied Sciences Journal, *4*(2) 300–307.

Niknam, T., Amiri, B., Olamaie, J., & Arefi, A. (2008). An efficient hybrid evolutionary optimization algorithm based on PSO and SA for clustering. Journal of Zhejiang University Science A. doi:10.1631/jzus.A0820196.

References

Raudenbush, S. W. (1997). Statistical analysis and optimal design for cluster randomized trials. *Psychological Methods, 2*(2), 173.

Roy, B. (1968). Classement et choix en présence de points de vue multiples. Revue française d'automatique, d'informatique et de recherche opérationnelle. Recherche opérationnelle, *2*(1), 57–75.

Roy, B. (1972). How outranking relation helps multiple criteria decision making. SEMA (Metra International), Direction Scientifique.

Roy, B., & Bouyssou, D. (1993). Aide multicritere a la decision: Methodes et cas. Economica, Paris.

Ruspini, E. R. (1969). A new approach to clustering. *Information Control, 19*, 22–32.

Ruspini, E. R. (1973). New experimental results in fuzzy clustering. *Information Sciences, 6*, 273–284.

Selim, S. Z., & Ismail, M. A. (1984). k-Means-type algorithms: A generalized convergence theorem and character- ization of local optimality. *IEEE Transactions on Pattern Analysis and Machine Intelligence, 6*(1), 81–87.

Shelokar, P. S., Jayaraman, V. K., & Kulkarni, B. D. (2004). An ant colony approach for clustering. *Analytica Chimica Acta, 509*(2), 187–195.

Siskos, J., & Zopounidis, C. (1987). The evaluation criteria of the venture capital investment activity. An interactive assessment. *European Journal of Operational Research, 31*, 304–313.

Smolinski, A., Walczak, B., & Einax, J. W. (2002). Hierarchical clustering extended with visual complements of environmental data set. *Chemometrics and Intelligent Laboratory Systems, 64*, 45–54.

Sneath, P. H., & Sokal, R. R. (1973). Numerical taxonomy. The principles and practice of numerical classification.

Steinbach, M., Ertoz, L., & Kumar, V. (2003). Challenges of clustering in high dimensional data. *University of Minnesota Supercomputing Institute Research Report, 213*, 1–33.

Stiakakis, E., & Sifaleras, A. (2013). Combining the priority rankings of DEA and AHP methodologies: a case study on an ICT industry. *International Journal of Data Analysis Techniques and Strategies, 5*(1), 101–114.

Struass, R. E. (1982). Statistical significance of species clusters in association analysis. *Ecology, 63*(3), 634–639.

Sung, C. S., & Jin, H. W. (2000). A tabu-search-based heuristic for clustering. *Pattern Recognition, 33*(5), 849–858.

Suzuki, R., & Shimodaira, H. (2006). Pvclust: an R package for assessing the uncertainty in hierarchical clustering. *Bioinformatics, 22*(12), 1540–1542.

Thorndike, R. L. (1953). Who belongs in a family? *Psychometrika, 18*, 267–276.

Vincke, J. P., & Brans, Ph. (1985). A preference ranking organization method. The PROMETHEE method for MCDM. *Management Science, 31*, 641–656.

Waller, N. G., Kaiser, H. A., Illian, J. B., & Manry, M. (1998). A comparison of the classification capabilities of the 1-dimensional kohonen neural network with two pratitioning and three hierarchical cluster analysis algorithms. *Psychometrika, 63*(1), 5–22.

Wiecek, M. M., Ehrgott, M., Fadel, G., & Figueira, J. R. (2008). Editorial: Multiple criteria decision making for engineering. *Omega, 36*, 337–339.

Zalik, K. R. (2008). An efficient k-means clustering algorithm. *Pattern Recognition Letters, 29*, 1385–1391.

Zopounidis, C. (1999). Multicriteria decision aid in financial management. *European Journal of Operational Research, 119*(2), 404–415.

Zopounidis, C. (2001). *Analysis of financing decisions using multiple criteria*. Thessaloniki: Anikoula Publications (pp. 67–85).

Zopounidis, C., & Doumpos, M. (1999). A multicriteria decision aid methodology for sorting decision problems: The case of financial distress. *Computational Economics, 14*(3), 197–218.

Zopounidis, C., & Doumpos, M. (2000). Building additive utilities for multi-group hierarchical discrimination: The MHDIS method. *Optimization Methods and Software, 14*(3), 219–240.

Zopounidis, C., & Doumpos, M. (2002). Multicriteria classification and sorting methods: A literature review. *European Journal of Operational Research, 138*(2), 229–246.

Zopounidis, C., Lemonakis, C., Andreopoulou, Z., & Koliouska, C. (2014). Agrotourism industry development through internet technologies: a multicriteria approach. *Journal of Euromarketing, 23*(4), 45–67.

Chapter 3
Applications in Various Agricultural, Food and Environmental Issues

In this section, we will apply four methodologies to data regarding agricultural, food and environmental issues. More specifically, we will use the following methods: the PROMETHEE II ranking method, the Hierarchical Cluster Analysis, the K-means analysis and the combination of PROMETHEE II method and a cluster generation method.

3.1 PROMETHEE II Method

Three examples will be presented and discussed in order to understand how PROMETHEE II method is used. The first example deals with skiing centers, the second example is about the agrotourism enterprises and the third one regards the aquaculture units. The above entities are ranked according to certain characteristics.

3.1.1 Ranking of Skiing Centers

The websites of 20 skiing centers were used for this example. Initially, qualitative analysis was performed in order to examine the type of common e-marketing criteria found in these skiing centers websites; then a quantitative analysis was carried out, in order to examine the presence or absence of these criteria/characteristics.

Various e-marketing services were introduced in the retrieved websites and 5 different criteria were identified and introduced in each website. Each e-marketing service constitutes a criteria/characteristic and it is finally attributed in a variable Xi (Table 3.1). Additionally, a 2-dimentional table was developed and was used in order to examine the existence of the criteria and evaluate the services of the websites. For that purpose, the values were attributed to variables X_1 to X_5, respectively.

Table 3.1 Variables attributed to criteria to be achieved by the skiing centers

Variable	E-marketing services to become criteria achieved by the ski centers website
X_1	Capacity of information provision to online visitors (weather forecast, accommodation, entertainment)
X_2	Interactivity and Online communication
X_3	Provision of advertising other companies through the website
X_4	Autonomous presence within the internet
X_5	Promotional activities, sales discount etc.

Variable X_1 refers to the capacity of information provision to the online visitors. The website provides up-to-date detailed local and theme based information, destination maps, generation of weather forecast warnings, enhanced navigational experience through web camera, snow cover maps and statistical summaries of snow. Variable X_2 represents the interactivity feature where the skiing center website provides online reservation functionalities, online booking system for accommodations, packages offered online, online rental of sports equipment, e-ticketing service and e-mail marketing. Also, the potential customers are able to chat with the employees of the skiing center for further information or with each other in a virtual community (social media). E-mail marketing is allowing the ski center to create the equivalent of impulse shopping at the check-out stand. The addition of direct mail to the marketing mix seems like a natural, given the resorts' mostly young and technology savvy audience. The resort solidifies its relationships by taking guests to a conversational, personal level online. Variable X_3 stands for the existence of advertisements for other local enterprises, such as restaurants, hotels, ski-equipment rentals, travel agency, etc. The autonomous internet presence (X_4) might include: access to the internet, e-mail address, website, listings in directories/search engines and other communication tools. The skiing centers that don't have an autonomous internet presence can be found through websites with general information about the skiing centers in Greece. Finally, variable X_5 represents the capability of sales discounts such as online exclusive coupons. It also represents the existence of other promotional activities. Sending last-minute e-mail invitations will contribute to the dramatic increase in the frequency of guest visits to the resort.

Whenever a criterion was achieved for a website the value 1 was attributed to the respective variable aiming at justifying the relative service within the evaluation of the website. The findings are presented in Table 3.2 and it was further analyzed the achievement of each service criteria in the sample websites.

Then, the total ranking of the websites was studied. The method that was used for the total ranking was the multicriteria analysis named PROMETHEE II. That method applies a linear form of service in this particular case, using the e-marketing services of the websites identified as criteria.

The net flow is the final number that is used for the comparison between the websites in order to obtain the ranking. The ten values (scenarios) range between

3.1 PROMETHEE II Method

Table 3.2 Data—skiing centers

S_C	X_1	X_2	X_3	X_4	X_5
S_C_1	1	1	1	1	1
S_C_2	0	0	0	0	0
S_C_3	0	0	0	0	0
S_C_4	1	1	1	1	1
S_C_5	1	1	0	1	1
S_C_6	0	0	0	0	0
S_C_7	1	1	1	1	1
S_C_8	1	1	0	1	0
S_C_9	1	1	1	1	1
S_C_10	0	0	0	0	0
S_C_11	1	1	1	1	1
S_C_12	0	0	0	0	0
S_C_13	0	0	0	0	0
S_C_14	0	0	0	0	0
S_C_15	0	0	0	0	0
S_C_16	0	0	0	0	0
S_C_17	1	1	0	1	1
S_C_18	0	0	0	0	0
S_C_19	1	1	1	1	1
S_C_20	1	1	1	1	1

0.25 and 2.5s with step 0.25s, where s is the standard deviation of all differences d for each criterion. In total, we take 500 net flow values for each website and find the website's average value. Each website with a higher net flow is considered superior in the final ranking (Tsekouropoulos et al. 2012b).

According to these findings (Table 3.3), the values estimated for total net flows ϕ present a great spectrum of values between +6,01 to −3,31 and that indicates a great difference concerning "superiority" between the first and the last case in the ranking of the enterprises' websites. The higher the net flow is, the better that respective alternative is (Euroconsultants 2011). As a result of the calculations made, we obtained the following ranking of the alternatives/cases studied: "SC_19" alternative ranked the first, "SC_20" alternative ranked the second, "SC_11" alternative ranked the third, "SC_9" alternative ranked the fourth, etc. From the study made, one notices that "SC_19" alternative ranked the first, and therefore we recommend the "SC_19" as the best solution. The rest skiing centers should definitely evolve and include innovative e-services to improve their websites and the website of "SC_19" should be their model in that process.

Table 3.3 Total net flow—skiing centers

	S_C	Net flow—linear
1	SC_19	6.011316254
2	SC_20	6.011316254
3	SC_11	4.77378838
4	SC_9	4.576232777
5	SC_7	4.485468572
6	SC_17	4.278536543
7	SC_4	4.235718711
8	SC_1	3.840607506
9	SC_5	3.065854414
10	SC_8	1.663159746
11	SC_18	−2.13015475
12	SC_12	−2.275516094
13	SC_13	−2.275516094
14	SC_14	−2.275516094
15	SC_15	−2.275516094
16	SC_16	−2.275516094
17	SC_10	−2.473071697
18	SC_6	−2.974974299
19	SC_2	−3.317891245
20	SC_3	−3.317891245

3.1.2 Ranking of Agrotourism Enterprises

The websites of 20 agrotourism enterprises were used for this example. There have been identified 13 different type of features introduced in each website aiming to promote agrotourism activities. These website features are described in Table 3.4 and they are attributed to 13 variables, $X_1, X_2, \ldots X_{13}$, that also act as criteria for the assessment of the website (Andreopoulou and Koutroumanidis 2009).

Variable X_1 refers to the detailed information on the agrotourism enterprise, the supported activities and services. Variable X_2 represents the existence of the current price list for the offered services, activities, products and special offers. Variable X_3 stands for the existence of contact information in the type of telephone, address, e-mail, etc. to enhance communication with the enterprise. Variable X_4 refers to features such as "local information" usually with texts, photo galleries and maps of the local area with tourism attractions and landmarks aiming to appeal the potential visitor. Variable X_5 represents the provided links to other similar entities, usually locally. Variable X_6 stands for information for other related sources of information concerning agrotourism, sustainable development, sustainable agriculture, development programs through the EU, such as LEADER+, local development agencies, etc. Variable X_7 refers to the feature about online reservation where website users can easily proceed and reserve their staying in the agotourism enterprise through reservation forms. Variable X_8 represents the online payment feature. Variable X_9

3.1 PROMETHEE II Method

Table 3.4 Variables attributed to criteria to be achieved by the agrotourism enterprises

Criteria	Main features of each
X_1	Information about products-services-activities
X_2	Current prices
X_3	Contact Information
X_4	Local information
X_5	Links to other companies etc.
X_6	Related sources of information
X_7	On line reservation (enabled with traditional ways of payment)
X_8	On line reservation (enabled with online payment)
X_9	On line communities (forums, chat rooms, guestbooks etc.)
X_{10}	Additional topics with information on different categories of interest
X_{11}	Code access
X_{12}	Third person advertisement
X_{13}	Personalization of the page, trace, safety

stands for the communication features for the users of the website such as forums, guestbook etc. where visitors can share their experience and communicate. Variable X_{11} stands for the existence of code access involving membership, while variable X_{12} stands for the third person advertisement. Finally, variable X_{13} refers to the personalization of the website through safety features. The findings are presented in Table 3.5 and it was further analyzed the achievement of each service criteria in the sample websites.

Then, the total ranking of the websites was studied. The method that was used for the total ranking was the multicriteria analysis named PROMETHEE II. That method applies a linear form of service in this particular case, using the e-services of the websites identified as criteria.

The net flow is the final number that is used for the comparison between the websites in order to obtain the ranking. The ten values (scenarios) range between 0.25 and 2.5s with step 0.25s, where s is the standard deviation of all differences d for each criterion. In total, we take 500 net flow values for each website and find the website's average value. Each website with a higher net flow is considered superior in the final ranking.

According to these findings (Table 3.6), the values estimated for total net flows φ present a great spectrum of values between +5,02 to −1,47 and that indicates a great difference concerning "superiority" between the first and the last case in the ranking of the enterprises' websites. The higher the net flow is, the better that respective alternative is (Euroconsultants 2011). As a result of the calculations made, we obtained the following ranking of the alternatives/cases studied: "A_E_12" alternative ranked the first, "A_E_16" alternative ranked the second, "A_E_17" alternative ranked the third, "A_E_1" alternative ranked the fourth, etc. From the study made, one notices that "A_E_12" alternative ranked the first, and therefore we recommend the "A_E_12" as the best solution. The rest agrotourism

Table 3.5 Data—agrotourism enterprises

	X_1	X_2	X_3	X_4	X_5	X_6	X_7	X_8	X_9	X_{10}	X_{11}	X_{12}	X_{13}
A_E_1	1	1	1	1	1	1	1	0	0	1	0	0	0
A_E_2	1	1	1	1	1	0	0	0	0	0	0	0	0
A_E_3	1	1	1	1	0	0	1	0	0	1	0	0	0
A_E_4	1	0	1	1	0	1	0	0	1	1	0	0	0
A_E_5	1	1	1	1	0	0	0	0	0	1	0	0	0
A_E_6	1	1	1	1	0	1	1	0	0	1	0	0	0
A_E_7	1	0	1	1	0	0	0	0	0	1	0	0	0
A_E_8	1	1	1	1	0	0	1	0	0	0	0	0	0
A_E_9	1	0	1	1	0	1	1	0	1	1	0	0	0
A_E_10	1	0	1	0	0	0	1	0	1	1	0	0	0
A_E_11	1	0	1	1	1	0	0	0	0	1	0	0	0
A_E_12	1	1	1	1	1	1	1	0	1	1	0	0	0
A_E_13	1	1	1	1	1	0	0	0	1	1	0	1	0
A_E_14	1	0	1	1	1	0	1	0	0	1	0	1	0
A_E_15	1	1	1	1	0	0	0	0	0	1	0	1	0
A_E_16	1	1	1	1	1	1	0	0	0	0	0	0	0
A_E_17	1	1	1	1	0	1	1	0	0	1	0	0	0
A_E_18	1	0	1	1	0	0	0	0	0	1	0	0	0
A_E_19	1	0	1	1	0	0	0	0	1	1	0	0	0
A_E_20	1	1	1	1	0	0	0	0	1	1	0	0	0

enterprises should definitely evolve and include innovative features to improve their websites aiming to promote their activities and the website of "A_E_12" should be their model in that process.

3.1.3 Ranking of Aquaculture Units

The websites of 20 aquaculture units were used for this example. There have been identified 13 different type of features introduced in each website aiming to promote their activities. There have been identified 13 different type of material introduced in each website aiming to promote e-commerce (Andreopoulou et al. 2009). These website features are described in Table 3.1 and they are attributed to 13 variables, $X_1, X_2, \ldots X_{13}$, that also act as criteria for the assessment of the website (Table 3.7).

Variable X_1 refers to the characteristic of information about the products and services that the enterprise offers to the possible e-customers. Variable X_2 stands for the characteristic of current prices lists for the products or services of the enterprise. Variable X_3 represents the characteristic of existence of promotional and informational material concerning the entity. That is an effort to strengthen the trust of

3.1 PROMETHEE II Method

Table 3.6 Total net flows—agrotourism enterprises

	A_E	Net flow—linear
1	A_E_12	5.024987044
2	A_E_16	3.860533613
3	A_E_17	3.85691232
4	A_E_1	3.723668743
5	A_E_6	2.743886309
6	A_E_14	2.221496152
7	A_E_13	2.140422601
8	A_E_9	1.54088064
9	A_E_8	1.509494744
10	A_E_3	1.224356759
11	A_E_2	0.976921664
12	A_E_20	0.949848853
13	A_E_15	0.669874628
14	A_E_11	0.560257107
15	A_E_5	−0.18530949
16	A_E_4	−0.337169793
17	A_E_18	−0.470366944
18	A_E_19	−0.470366944
19	A_E_10	−1.345223562
20	A_E_7	−1.479938259

Table 3.7 Variables attributed to criteria to be achieved by the aquaculture units

Variable	Criteria achieved by the website
X_1	Technical Information about the products/services
X_2	Information about the current prices for the products/services
X_3	Information about the carrier/owner
X_4	Information about transaction policies and local information
X_5	Links to other relative companies, organisations, carriers, etc.
X_6	Links to other relative elements and sources of information
X_7	Enabled online transactions with traditional ways of payment
X_8	Enabled online transactions with also enabled online payment
X_9	On line communities such as forums and chat rooms
X_{10}	Additional topics with information on different categories of interest
X_{11}	Code access
X_{12}	Third person advertisement
X_{13}	Personalization of the page, trace, safety

the potential client to the enterprise, a vital factor for on-line sales. Variable X_4 stands for the provision of information on transaction accomplishment and policy of the enterprise. Variable X_5 refers to the provided links to relative organizations, enterprises and carriers, while variable X_6 refers to the provided links for sources of information in the Internet. Variable X_7 represents the on line transactions with traditional payment, as in these cases it is usually enabled only the on line communication by email. Usually, just a simple telephone number is available for contact and orders. Variable X_8 stands for the economic transaction can be fulfilled through the website. Variable X_9 refers to the on line communities such as forums and chat rooms. Variable X_{10} refers to the identified topics with informational material in the website for various issues and different categories of interest, such as information on the area, environmental issues, cooking proposals on fish etc. Variable X_{11} represents the characteristic of a website area where access is allowed only for members through codes and passwords (code access). Code access is usually performed by e-catalogs that promote aquaculture units. Variable X_{12} represents the third person advertisement. Variable X_{13} stands for the possibility for personalization of the website. The findings are presented in Table 3.8 and it was further analyzed the achievement of each service criteria in the sample websites.

Then, the total ranking of the websites was studied. The method that was used for the total ranking was the multicriteria analysis named PROMETHEE II. That method applies a linear form of service in this particular case, using the e-services of the websites identified as criteria.

Table 3.8 Data—aquaculture units

	X_1	X_2	X_3	X_4	X_5	X_6	X_7	X_8	X_9	X_{10}	X_{11}	X_{12}	X_{13}
A_U_1	1	0	1	1	1	1	0	0	0	0	0	1	0
A_U_2	1	0	1	1	1	1	0	0	0	1	0	0	0
A_U_3	1	0	1	0	0	1	0	0	1	0	0	0	0
A_U_4	1	0	1	1	0	0	0	0	0	1	0	0	0
A_U_5	1	1	1	1	0	0	1	1	1	1	0	0	0
A_U_6	1	0	1	1	1	1	0	0	0	1	0	1	0
A_U_7	1	0	1	0	1	1	0	0	0	1	0	1	0
A_U_8	1	0	0	0	0	0	0	0	0	0	0	0	0
A_U_9	1	0	1	0	1	0	0	0	0	0	0	0	0
A_U_10	1	1	1	0	1	1	0	0	0	1	0	0	0
A_U_11	1	1	1	1	0	1	0	0	0	1	0	0	0
A_U_12	1	0	1	1	1	1	0	0	0	1	0	0	0
A_U_13	1	0	1	1	1	1	1	0	0	1	0	0	0
A_U_14	1	0	1	1	1	1	1	0	1	0	0	0	0
A_U_15	1	1	1	1	1	1	1	0	1	0	1	0	0
A_U_16	1	0	1	0	1	1	0	0	0	1	1	0	0
A_U_17	1	0	0	0	1	1	0	0	0	1	1	0	0
A_U_18	0	0	1	0	1	1	0	0	0	1	1	0	0
A_U_19	0	0	1	0	1	0	0	0	0	1	1	0	0
A_U_20	0	0	1	0	1	0	0	0	0	1	1	0	0

3.1 PROMETHEE II Method

Table 3.9 Total net flows—aquaculture units

	A_U	Net flow—linear
1	A_U_15	6.044486231
2	A_U_13	4.512561447
3	A_U_14	4.512561447
4	A_U_12	2.753250944
5	A_U_10	2.422382976
6	A_U_11	2.286483728
7	A_U_6	2.262072354
8	A_U_1	1.748435007
9	A_U_2	1.748435007
10	A_U_5	1.675121182
11	A_U_16	1.629003434
12	A_U_18	1.043314172
13	A_U_7	0.797987996
14	A_U_17	0.593871301
15	A_U_19	−0.256498173
16	A_U_20	−0.256498173
17	A_U_9	−0.663147858
18	A_U_4	−1.508068521
19	A_U_3	−1.519851029
20	A_U_8	−3.929082148

The net flow is the final number that is used for the comparison between the websites in order to obtain the ranking. The ten values (scenarios) range between 0.25 and 2.5s with step 0.25s, where s is the standard deviation of all differences d for each criterion. In total, we take 500 net flow values for each website and find the website's average value. Each website with a higher net flow is considered superior in the final ranking.

According to these findings (Table 3.9), the values estimated for total net flows ϕ present a great spectrum of values between +6,04 to −3,92 and that indicates a great difference concerning "superiority" between the first and the last case in the ranking of the websites. The higher the net flow is, the better that respective alternative is (Euroconsultants 2011). As a result of the calculations made, we obtained the following ranking of the alternatives/cases studied: "A_U_15" alternative ranked the first, "A_U_13" alternative ranked the second, "A_U_14" alternative ranked the third, "A_U_12" alternative ranked the fourth, etc. From the study made, one notices that "A_U_15" alternative ranked the first, and therefore we recommend the "A_U_15" as the best solution. The aquaculture units should definitely evolve and include innovative e-commerce features to improve their websites aiming to promote their activities and the website of "A_U_15" should be their model in that process.

3.2 Hierarchical Cluster Analysis

Two examples will be presented and discussed in order to understand how hierarchical cluster analysis is implemented for the classification of enterprises. The first example deals with wood enterprises and the second one is about the enterprises which promote the Renewable Energy Sources. The above entities are classified according to certain characteristics.

3.2.1 Classification of Wood Enterprises

The websites of 20 wood enterprises were used for this example. Initially, qualitative analysis was performed in order to examine the type of common characteristics found in these wood enterprises websites; then a quantitative analysis was carried out, in order to examine the presence or absence of these criteria/characteristics (Andreopoulou et al. 2009).

Various materials were introduced in the retrieved websites aiming to promote the products and enterprises. 13 different types of material were identified and introduced in each website that aim to promote e-commerce. Each e-commerce characteristic constitutes a criteria/characteristic and it is finally attributed in a variable X_i (Table 3.10). Additionally, a 2-dimentional table was developed and was used in order to examine the existence of the criteria and evaluate the services of the websites. For that purpose, the values were attributed to variables X_1 to X_{13}, respectively.

Variable X_1 refers to the characteristic of information about the products and services that the enterprise offers to the possible e-customers. Variable X_2 stands for

Table 3.10 Variables attributed to criteria to be achieved by the wood enterprises

Variable	E-commerce characteristics
X_1	Information about the products/services
X_2	Information about the current prices of the products/services
X_3	Information about the carrier/owner
X_4	Information about the transaction fulfilment and policy
X_5	Links to other relative enterprises, organizations, carriers, etc.
X_6	Links to other relative issues and sources of information
X_7	Enabled online transactions with traditional ways of payment
X_8	Enabled online transactions with online payment enabled
X_9	Online communities such as forums and chat rooms
X_{10}	Additional topics with information on different categories of interest
X_{11}	Code access
X_{12}	Third-person advertisement
X_{13}	Personalization of the page

the characteristic of current prices lists for the products or services of the enterprise. Variable X_3 represents the characteristic of existence of promotional and informational material concerning the entity. That is an effort to strengthen the trust of the potential client to the enterprise, a vital factor for on-line sales. Variable X_4 stands for the provision of information on transaction accomplishment and policy of the enterprise. Variable X_5 refers to the provided links to relative organizations, enterprises and carriers, while variable X_6 refers to the provided links for sources of information in the Internet. Variable X_7 represents the on line transactions with traditional payment, as in these cases it is usually enabled only the on line communication by email. Usually, just a simple telephone number is available for contact and orders. Variable X_8 stands for the economic transaction can be fulfilled through the website. Variable X_9 refers to the on line communities such as forums and chat rooms. Variable X_{10} refers to the identified topics with informational material in the website for various issues and different categories of interest, such as information on the area, environmental issues, cooking proposals on fish etc. Variable X_{11} represents the characteristic of a website area where access is allowed only for members through codes and passwords (code access). Code access is usually performed by e-catalogs that promote aquaculture units. Variable X_{12} represents the third person advertisement. Variable X_{13} stands for the possibility for personalization of the website.

Whenever a criterion was achieved for a website the value 1 was attributed to the respective variable aiming at justifying the relative service within the evaluation of the website. The findings are presented in Table 3.11 and it was further analyzed the achievement of each service criteria in the sample websites.

Hierarchical cluster analysis was used to identify homogenous groups of wood enterprises that have similar e-services characteristics but are distinctively different from other wood enterprise segments. In Table 3.12, the case processing summary is presented which lists the number of valid cases and the number of missing cases.

The proximity matrix (Table 3.13) provides the actual distances, which reveals the similarities computed for any pair of observations and variables.

The agglomeration schedule (Table 3.14) details how observations are clustered together at each stage of the cluster analysis. When clusters or cases are joined, they are subsequently labeled with the smaller of the two cluster numbers. The Coefficients column indicates the distance between the two clusters (or cases) joined at each stage. For a good cluster solution, you will see a sudden jump in the distance coefficient (or a sudden drop in the similarity coefficient) as you read down the table (Stage 17: Coefficient 3.000, Stage 18: Coefficient 4.244). The stage before the sudden change indicates the optimal stopping point for merging clusters (Stage 17). For this example, 3 clusters remain after Stage 17. So, we should consider using a 3-cluster solution.

The cluster membership (Table 3.15) provides detailed group structure after classification.

This plot (Fig. 3.1) gives a graphic representation of how the cases are joined at each stage of the analysis. Each white bar represents a boundary between clusters. At each stage, two clusters are joined, and so the white bar separating the joined

Table 3.11 Data—wood enterprises

W_E	X_1	X_2	X_3	X_4	X_5	X_6	X_7	X_8	X_9	X_{10}	X_{11}	X_{12}	X_{13}
W_E_1	1	0	1	1	0	1	0	0	0	1	0	0	1
W_E_2	1	0	1	0	0	0	0	0	0	1	0	0	0
W_E_3	1	0	1	1	0	0	0	0	0	0	0	0	0
W_E_4	1	0	1	1	0	0	0	0	0	0	0	0	0
W_E_5	1	0	1	1	0	0	1	0	1	0	1	0	1
W_E_6	1	0	1	1	0	0	0	0	0	0	0	0	1
W_E_7	1	0	1	1	0	0	0	0	0	0	0	0	0
W_E_8	1	0	1	1	0	0	0	0	0	0	0	0	0
W_E_9	1	0	1	1	0	1	0	0	1	0	0	0	1
W_E_10	1	0	1	1	0	1	0	1	1	0	0	0	1
W_E_11	1	0	0	1	0	0	1	0	1	1	1	0	1
W_E_12	1	0	0	0	0	0	0	0	0	0	0	0	0
W_E_13	1	0	1	1	0	0	1	0	1	0	0	0	0
W_E_14	1	0	1	0	1	0	1	0	0	0	0	0	0
W_E_15	1	0	1	1	0	0	1	0	0	0	0	0	0
W_E_16	1	0	1	1	1	0	0	0	0	0	0	0	0
W_E_17	1	0	1	1	1	0	1	0	0	1	0	0	0
W_E_18	1	0	1	0	0	0	0	0	0	0	0	0	0
W_E_19	1	0	1	1	1	0	0	0	0	0	0	0	0
W_E_20	1	0	1	1	1	0	1	0	0	0	0	0	0

Table 3.12 Case processing summary—wood enterprises

Cases					
Valid		Missing		Total	
N	Percent	N	Percent	N	Percent
20	100.0	0	0	20	100.0

Table 3.13 Proximity matrix—wood enterprises (part)

	Squared euclidean distance								
Case	1	2	3	4	5	6	7	8	9
1	0	3.000	3.000	3.000	5.000	2.000	3.000	3.000	0
2	3.000	0	2.000	2.000	6.000	3.000	2.000	2.000	3.000
3	3.000	2.000	0	0	4.000	1.000	0	0	3.000
4	3.000	2.000	0	0	4.000	1.000	0	0	3.000
5	5.000	6.000	4.000	4.000	0	3.000	4.000	4.000	5.000
6	2.000	3.000	1.000	1.000	3.000	0	1.000	1.000	2.000
7	3.000	2.000	0	0	4.000	1.000	0	0	3.000
8	3.000	2.000	0	0	4.000	1.000	0	0	3.000
9	0	3.000	3.000	3.000	5.000	2.000	3.000	3.000	0

3.2 Hierarchical Cluster Analysis

Table 3.14 Agglomeration schedule—wood enterprises

Stage	Cluster combined		Coefficients	Stage cluster first appears		Next stage
	Cluster 1	Cluster 2		Cluster 1	Cluster 2	
1	16	19	0	0	0	8
2	1	9	0	0	0	17
3	7	8	0	0	0	4
4	3	7	0	0	3	5
5	3	4	0	4	0	8
6	14	20	1.000	0	0	11
7	12	18	1.000	0	0	12
8	3	16	1.000	5	1	10
9	13	15	1.000	0	0	13
10	3	6	1.333	8	0	13
11	14	17	1.500	6	0	16
12	2	12	1.500	0	7	15
13	3	13	1.929	10	9	15
14	5	11	2.000	0	0	19
15	2	3	2.333	12	13	16
16	2	14	2.694	15	11	18
17	1	10	3.000	2	0	18
18	1	2	4.244	17	16	19
19	1	5	5.111	18	14	0

Table 3.15 Cluster membership—wood enterprises

W_E	2 Clusters	W_E	2 Clusters
W_E_1	1	W_E_11	2
W_E_2	1	W_E_12	1
W_E_3	1	W_E_13	1
W_E_4	1	W_E_14	1
W_E_5	2	W_E_15	1
W_E_6	1	W_E_16	1
W_E_7	1	W_E_17	1
W_E_8	1	W_E_18	1
W_E_9	1	W_E_19	1
W_E_10	1	W_E_20	1

clusters ends. Within a row, each contiguous black band indicates cases grouped as a cluster.

The results of the cluster analysis is a dendrogram (Fig. 3.2) with $19(n-1)$ nodes. The horizontal axis of the dendrogram represents the distance or dissimilarity between clusters. The vertical axis represents the cases and clusters. Each joining (fusion) of two clusters is represented on the graph by the splitting of a

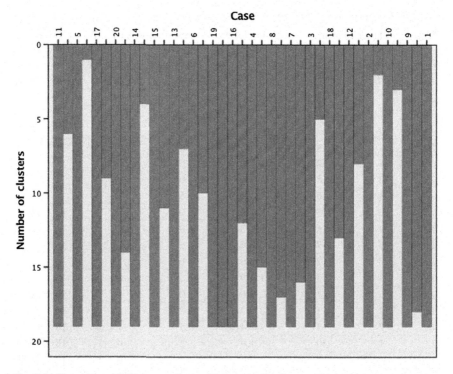

Fig. 3.1 Clustering with "Between Groups Linkage method"—wood enterprises

horizontal line into two horizontal lines. The horizontal position of the split, shown by the short vertical bar, gives the distance between the two clusters. As longer the horizontal line is, more stable the previous clustering tends to be.

The number of clusters the will be formed at a particular Cluster Cutoff value may be quickly determined from this plot by drawing a vertical line at that value and counting the number of lines that the vertical line intersects. For example, we can see that if we draw a vertical line at the value 1.1, six clusters will result. One cluster will contain nine objects, two clusters each will contain three objects, two clusters each will contain two objects and one cluster will contain only one object. Cluster analysis consists a tool to create a typology for websites characteristics of wood enterprises.

3.2.2 Typology of RES Enterprises

The websites of 20 enterprises that promote Renewable Energy Sources (RES) were used for this example. Initially, qualitative analysis was performed in order to examine the type of common features found in these RES enterprises websites; then

3.2 Hierarchical Cluster Analysis

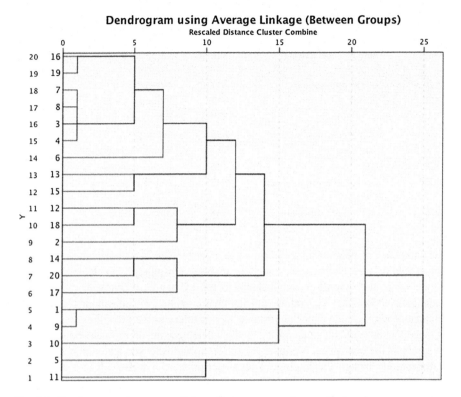

Fig. 3.2 Dendrogram using average linkage (between groups)—wood enterprises

a quantitative analysis was carried out, in order to examine the presence or absence of these criteria/characteristics.

Various features were introduced in the retrieved websites aiming to represent the internet adoption of these RES enterprises. 9 different types of features were identified and introduced in each enterprise website. Each feature constitutes a criteria/characteristic and it is finally attributed in a variable X_i (Table 3.16). Additionally, a 2-dimentional table was developed and was used in order to examine the existence of the criteria and evaluate the services of the websites. For that purpose, the values were attributed to variables X_1 to X_{18}, respectively.

The first variable refers to the ability to view the content of an internet presence in more than two languages (Greek and English). Variable X2 is associated with the provision of an interactive digital map for the orientation of website visitors–customers. Variable X_3 stands for the provision of a pricelist and variable X_4 is associated with the provision of useful links to other relevant organizations or enterprises. Variable X_5 refers to the participation of the enterprise in social media, while Variable X_6 represents the characteristic of a website area where access is allowed only for members through codes and passwords (code access). Variable X_7 refers to the tab of the Frequently Asked Questions (FAQ) and variable X_8 refers to

Table 3.16 Variables attributed to criteria to be achieved by the RES enterprises

Variable	Features	Variable	Features
X1	Multilanguage website	X10	Information about enterprise activities
X2	Google map service	X11	Newsletter service
X3	Pricelist	X12	Projects
X4	Useful links	X13	RSS feeds
X5	Social media profile	X14	Sitemap
X6	Code access	X15	Website search engine
X7	FAQ	X16	Online communication form
X8	General information about energy	X17	Online order form
X9	Online profit calculator	X18	Special offers

the provision of general information about energy. The ninth variable refers to the provision of any audiovisual material, such as photographs, videos and virtual tours, while the seventh variable refers to the existence of live web online profit calculator. Variable X_{10} refers to the characteristic of information about the enterprise activities. As for the variable X_{11}, it refers to the provision of newsletter service, while variable X_{12} refers to the characteristic of information about the projects of the enterprise. Variable X_{13} refers to the RSS service, which distributes information through the Internet. The subscription to a website RSS removes the need of manually checking the website for new content because the users' browser regularly monitors the internet presence and informs the users of any updates. Variable X_{14} refers to the provision of a sitemap for an overview of the website content and variable X_{15} represents the provision of website search engine. Variable X_{16} refers to the provision of online communication form and variable X_{17} refers to the provision of online order form. Last variable (X_{18}) represents the provision of special offers of the enterprise to its possible customers.

Whenever a criterion was achieved for a website the value 1 was attributed to the respective variable aiming at justifying the relative service within the evaluation of the website. As for variable X_1, the number of the available languages is inserted. The findings are presented in Table 3.17 and it was further analysed the achievement of each service criteria in the sample websites.

Hierarchical cluster analysis was used to identify homogenous groups of RES enterprises that have similar internet adoption characteristics but are distinctively different from other RES enterprise segments. In Table 3.18, the case processing summary is presented which lists the number of valid cases and the number of missing cases.

The proximity matrix (Table 3.19) provides the actual distances, which reveals the similarities computed for any pair of observations and variables.

The agglomeration schedule (Table 3.20) details how observations are clustered together at each stage of the cluster analysis. When clusters or cases are joined, they are subsequently labeled with the smaller of the two cluster numbers. The

3.2 Hierarchical Cluster Analysis

Table 3.17 Data—RES enterprises

RES	X_1	X_2	X_3	X_4	X_5	X_6	...	X_{13}	X_{14}	X_{15}	X_{16}	X_{17}	X_{18}
RES_1	1	1	0	0	3	0	...	0	0	0	1	0	0
RES_2	1	1	1	1	1	0	...	0	0	0	0	1	0
RES_3	1	1	1	1	1	0	...	0	0	0	0	1	0
RES_4	1	1	0	1	0	0	...	0	0	1	1	0	0
RES_5	2	1	0	0	4	0	...	1	1	0	1	0	0
RES_6	2	1	0	0	3	1	...	0	0	1	1	0	0
RES_7	2	1	0	0	0	0	...	0	0	1	1	0	0
RES_8	1	1	0	1	0	0	...	0	1	1	1	0	1
RES_9	2	1	0	1	0	1	...	0	0	1	1	0	0
RES_10	3	1	0	0	1	1	...	0	0	0	1	0	0
RES_11	1	1	0	0	1	0	...	1	1	1	1	0	0
RES_12	2	0	1	0	0	0	...	0	1	1	1	1	1
RES_13	1	1	0	0	1	0	...	0	0	1	1	0	0
RES_14	1	0	0	0	0	0	...	0	0	0	1	0	0
RES_15	1	1	0	0	0	0	...	0	0	0	1	0	0
RES_16	2	0	1	0	0	1	...	0	1	1	1	1	1
RES_17	1	0	0	1	0	0	...	0	0	0	1	0	0
RES_18	2	1	1	1	0	0	...	0	0	0	1	0	1
RES_19	3	1	0	1	0	1	...	0	0	0	1	0	1
RES_20	2	1	0	1	2	0	...	0	1	0	1	0	0

Table 3.18 Case processing summary—RES enterprises

Cases						
Valid		Missing		Total		
N	Percent	N	Percent	N	Percent	
20	100.0	0	0	20	100.0	

Coefficients column indicates the distance between the two clusters (or cases) joined at each stage. For a good cluster solution, you will see a sudden jump in the distance coefficient (or a sudden drop in the similarity coefficient) as you read down the table (Stage 18: Coefficient 9.786, Stage 19: Coefficient 14.156). The stage before the sudden change indicates the optimal stopping point for merging clusters (Stage 18). For this example, 2 clusters remain after Stage 18. So, we should consider using a 2-cluster solution.

The cluster membership (Table 3.21) provides detailed group structure after classification.

This plot (Fig. 3.3) gives a graphic representation of how the cases are joined at each stage of the analysis. Each white bar represents a boundary between clusters. At each stage, two clusters are joined, and so the white bar separating the joined clusters ends. Within a row, each contiguous black band indicates cases grouped as a cluster.

Table 3.19 Proximity matrix—RES enterprises (part)

Case	Squared euclidean distance								
	1	2	3	4	5	6	7	8	9
1	0	11.00	11.00	14.00	5.00	6.00	14.00	16.00	15.00
2	11.00	0	0	7.00	20.00	15.00	11.00	9.00	10.00
3	11.00	0	0	7.00	20.00	15.00	11.00	9.00	10.00
4	14.00	7.00	7.00	0	23.00	14.00	4.00	2.00	3.00
5	5.00	20.00	20.00	23.00	0	7.00	21.00	23.00	22.00
6	6.00	15.00	15.00	14.00	7.00	0	10.00	16.00	13.00
7	14.00	11.00	11.00	4.00	21.00	10.00	0	6.00	5.00
8	16.00	9.00	9.00	2.00	23.00	16.00	6.00	0	5.00
9	15.00	10.00	10.00	3.00	22.00	13.00	5.00	5.00	0

Table 3.20 Agglomeration schedule—RES enterprises

Stage	Cluster combined		Coefficients	Stage cluster first appears		Next stage
	Cluster 1	Cluster 2		Cluster 1	Cluster 2	
1	2	3	0	0	0	16
2	4	15	2.000	0	0	3
3	4	17	3.000	2	0	4
4	4	8	3.667	3	0	6
5	12	16	4.000	0	0	17
6	4	18	4.250	4	0	7
7	4	9	4.800	6	0	12
8	10	19	5.000	0	0	18
9	7	14	5.000	0	0	12
10	11	13	5.000	0	0	13
11	1	5	5.000	0	0	14
12	4	7	5.333	7	9	13
13	4	11	6.000	12	10	16
14	1	6	6.500	11	0	15
15	1	20	7.333	14	0	19
16	2	4	8.400	1	13	17
17	2	12	9.667	16	5	18
18	2	10	9.786	17	8	19
19	1	2	14.156	15	18	0

The result of the cluster analysis is a dendrogram (Fig. 3.4) with 19(n − 1) nodes. The horizontal axis of the dendrogram represents the distance or dissimilarity between clusters. The vertical axis represents the cases and clusters. Each joining (fusion) of two clusters is represented on the graph by the splitting of a horizontal line into two horizontal lines. The horizontal position of the split, shown

3.2 Hierarchical Cluster Analysis

Table 3.21 Cluster membership—RES enterprises

RES	2 Clusters	RES	2 Clusters
RES_1	1	RES_11	2
RES_2	2	RES_12	2
RES_3	2	RES_13	2
RES_4	2	RES_14	2
RES_5	1	RES_15	2
RES_6	1	RES_16	2
RES_7	2	RES_17	2
RES_8	2	RES_18	2
RES_9	2	RES_19	2
RES_10	2	RES_20	1

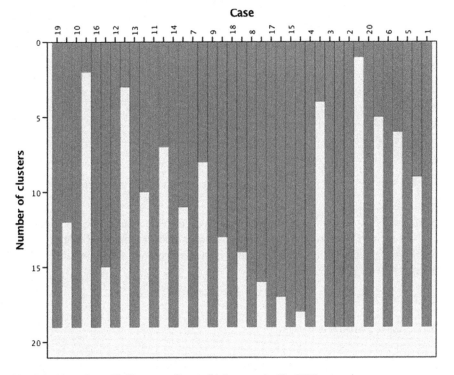

Fig. 3.3 Clustering with "Between Groups Linkage method"—RES enterprises

by the short vertical bar, gives the distance between the two clusters. As longer the horizontal line is, more stable the previous clustering tends to be.

The number of clusters the will be formed at a particular Cluster Cutoff value may be quickly determined from this plot by drawing a vertical line at that value and counting the number of lines that the vertical line intersects. For example, we can see that if we draw a vertical line at the value 1.1, eight clusters will result. One

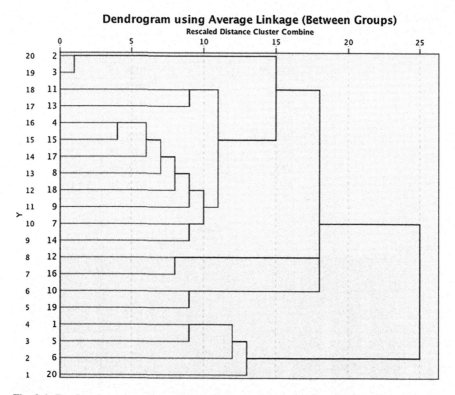

Fig. 3.4 Dendrogram using average linkage (between groups)—RES enterprises

cluster will contain eight objects, five clusters each will contain two objects and two clusters each will contain only one object. Cluster analysis consists a tool to create a typology for websites features of enterprises that promote the Renewable Energy Sources.

3.3 K-Means Analysis

Three examples will be presented and discussed in order to understand how K-means analysis is applied for the classification of government agencies, prefectures and agrifood enterprises. These entities are classified according to certain characteristics.

3.3 K-Means Analysis

3.3.1 Classification of Government Agencies in National Parks

The websites of 20 government agencies in the 10 Greek National Parks were used for this example. Initially, qualitative analysis was performed in order to examine the type of common characteristics found in these government agencies websites; then a quantitative analysis was carried out, in order to examine the presence or absence of these criteria/characteristics (Koliouska et al. 2015).

Various e-government services were introduced in the retrieved websites. 30 different types of e-government services were identified and introduced in each website. Each e-government service constitutes a criteria/characteristic and it is finally attributed in a variable X_i (Table 3.22). Additionally, a 2-dimentional table was developed and was used in order to examine the existence of the criteria and evaluate the services of the websites. For that purpose, the values were attributed to variables X_1 to X_{30}, respectively.

The first variable refers to the ability to view the content of an internet presence in more than two languages (Greek and English). Variable X_2 refers to the provision of detailed information about the products or services of the enterprise and the third one refers to the provision of information about the enterprise or organization (ownership, mailing address, telephone number and email address) that facilitate the

Table 3.22 Variables attributed to criteria to be achieved by the government agencies in National Parks

Variable	Feature	Variable	Feature
X_1	Two or more languages	X_{16}	Links to other companies etc.
X_2	Information about products-services-activities	X_{17}	Various topics of interest
X_3	Contact information	X_{18}	Downloadable files
X_4	Local information	X_{19}	Calendar application
X_5	Digital map	X_{20}	Event calendar application
X_6	Audiovisual material	X_{21}	Celebration calendar application
X_7	Live web camera	X_{22}	Social media sharing
X_8	Search engine	X_{23}	Social media profile
X_9	Sitemap	X_{24}	Forum
X_{10}	Updated enterprise information	X_{25}	Related sources of information
X_{11}	Online survey	X_{26}	Third person advertisement
X_{12}	Online communication form	X_{27}	Newsletter
X_{13}	Weather forecast	X_{28}	RSS
X_{14}	Website visitor tracker	X_{29}	Code access
X_{15}	Frequently Asked Questions (FAQ)	X_{30}	Personalization of the page, trace, safety

communication between the customers and the enterprises. As for the variable X_4, it refers to the provision of information about the local area.

Variable X_5 is associated with the provision of an interactive digital map for the orientation of website visitors–customers. The sixth variable refers to the provision of any audiovisual material, such as photographs, videos and virtual tours, while the seventh variable refers to the existence of live web camera. Variable X_8 represents the provision of web search engine. Variable X_9 refers to the provision of a sitemap for an overview of the website content. The tenth variable is associated with the continuous updating of the internet presence related to the activities of the enterprise. As for the variable X_{11}, it refers to the provision of the application for online polls to identify the users' trends. Variable X_{12} refers to the provision of online communication form. Variable X_{13} is associated with the provision of weather forecast application, while variable X_{14} refers to the website visitor tracker. Variable X_{15} refers to the tab of the Frequently Asked Questions (FAQ) and variable X_{16} is associated with the provision of useful links to other relevant organizations or enterprises. Variable X_{17} refers to the provision of information on different topics, while variable X_{18} refers to the ability to download some useful files. Variable X_{19} is associated with the clock and calendar application, variable X_{20} is associated with the local events calendar and variable X_{21} with the celebration calendar application integration. Variable X_{22} refers to the provision of sharing the internet presence through users' account in social networks, while variable X_{23} refers to the participation of the enterprise in social media. Variable X_{24} is associated with the provision of an online interactive community through the internet presence and variable X_{25} is associated with the available information on related topics. Variable X_{26} refers to the third person advertisement. As for the variable X_{27}, it refers to the provision of newsletter service. Variable X_{29} refers to the ability to create a user account. Variable X_{30} refers to the ability to be personalized the website by the registered users, while variable X_{28} refers to the RSS service, which distributes information through the Internet. The subscription to a website RSS removes the need of manually checking the website for new content because the users' browser regularly monitors the internet presence and informs the users of any updates.

Whenever a criterion was achieved for a website the value 1 was attributed to the respective variable aiming at justifying the relative service within the evaluation of the website. The findings are presented in Table 3.23 and it was further analyzed the achievement of each service criteria in the sample websites.

K-means analysis was used to identify homogenous groups of government agencies in National Parks that have similar e-government services characteristics but are distinctively different from other wood enterprise segments. The initial cluster centers are presented in Table 3.24. They are vectors with their values based on the 30 variables, which refer to the two clusters. These 2 clusters are at maximum index distance from each other.

In Table 3.25 we can see the number of the iterations and the changes in the cluster centers. In the second iteration the process of redistribution of the units stops because there are no changes in the cluster centers.

3.3 K-Means Analysis

Table 3.23 Data—government agencies in National Parks

G_A	X_1	X_2	X_3	X_4	X_5	X_6	...	X_{25}	X_{26}	X_{27}	X_{28}	X_{29}	X_{30}
G_A_1	2	0	1	0	0	1	...	1	0	1	0	1	0
G_A_2	1	0	0	0	1	0	...	1	0	1	0	0	1
G_A_3	1	1	1	0	1	1	...	1	0	1	0	1	0
G_A_4	1	0	0	0	0	0	...	1	0	1	0	0	1
G_A_5	5	0	1	0	1	1	...	1	0	1	0	1	0
G_A_6	1	0	1	0	1	0	...	1	1	1	1	0	0
G_A_7	1	0	1	1	1	0	...	1	1	1	1	0	1
G_A_8	2	0	1	0	0	0	...	1	0	0	0	1	0
G_A_9	5	0	1	0	1	1	...	1	1	1	1	1	0
G_A_10	1	0	1	0	0	0	...	1	1	1	1	0	0
G_A_11	2	0	1	0	1	1	...	1	0	1	0	0	0
G_A_12	1	0	1	0	1	1	...	1	0	1	0	0	0
G_A_13	8	0	1	0	1	1	...	1	0	1	0	0	0
G_A_14	2	0	1	0	1	1	...	1	1	1	1	0	0
G_A_15	3	1	1	0	1	1	...	1	1	1	1	1	0
G_A_16	2	1	1	0	1	0	...	1	0	1	0	0	1
G_A_17	2	1	1	0	1	0	...	1	0	1	0	0	0
G_A_18	1	0	1	0	1	0	...	1	0	1	1	0	0
G_A_19	1	1	1	1	1	0	...	1	0	1	0	0	0
G_A_20	1	0	1	0	1	0	...	1	1	1	0	0	1

Table 3.24 Initial cluster centers—government agencies in National

Variables	Cluster		Variables	Cluster	
	1	2		1	2
X_1	8	1	X_{16}	1	0
X_2	0	1	X_{17}	2	0
X_3	1	1	X_{18}	0	0
X_4	0	1	X_{19}	0	0
X_5	1	1	X_{20}	0	0
X_6	1	0	X_{21}	1	1
X_7	1	1	X_{22}	1	1
X_8	1	0	X_{23}	1	1
X_9	0	0	X_{24}	1	1
X_{10}	1	0	X_{25}	1	1
X_{11}	0	1	X_{26}	0	0
X_{12}	0	0	X_{27}	1	1
X_{13}	0	1	X_{28}	0	0
X_{14}	0	0	X_{29}	0	0
X_{15}	0	0	X_{30}	0	0

Table 3.25 Iteration history—government agencies in National

Iteration	Change in cluster centers	
	1	2
1	2.809	1.865
2	0	0

Table 3.26 Cluster membership—government agencies in National

Case Number	Cluster	Distance	Case Number	Cluster	Distance
G_A_1	2	1.751	G_A_11	2	2.390
G_A_2	2	1.665	G_A_12	2	1.912
G_A_3	2	1.700	G_A_13	1	2.809
G_A_4	2	1.896	G_A_14	2	1.593
G_A_5	1	1.700	G_A_15	2	2.669
G_A_6	2	1.682	G_A_16	2	2.636
G_A_7	2	2.157	G_A_17	2	2.031
G_A_8	2	2.290	G_A_18	2	1.375
G_A_9	1	2.134	G_A_19	2	1.865
G_A_10	2	1.849	G_A_20	2	1.987

The results are summarized in Table 3.26, i.e. which cluster each unit belongs to and the new cluster centers. The first cluster is formed by the cases G_A_5, G_A_9, G_A_13, G_A_14, G_A_15, G_A_16, G_A_17, G_A_18, G_A_19, G_A_20 and the second by G_A_1, G_A_2, G_A_3, G_A_4, G_A_6, G_A_7, G_A_8, G_A_10, G_A_11, G_A_12.

In Table 3.27 we can see the final cluster centers, and in Table 3.28—the distance between the final cluster centers.

Table 3.29 presents data for the number of units in each cluster as well as their total number and missing units (if there are any).

3.3.2 Typology of Prefectures According to Agricultural Exploitations

For this example, we have used the distribution of the areas (stremmas) of 50 Prefectures. Initially, qualitative analysis was performed in order to examine the type of management and marketing criteria found in these agrifood enterprises websites; then a quantitative analysis was carried out, in order to examine the presence or absence of these criteria/characteristics (Tsekouropoulos et al. 2012a).

Various management and marketing criteria were introduced in the retrieved corporate websites and different criteria were identified and introduced in each website. Each management and marketing function constitutes a criterion and it is

3.3 K-Means Analysis

Table 3.27 Final cluster centers—government agencies in National

Variables	Cluster		Variables	Cluster	
	1	2		1	2
X_1	6	1	X_{16}	0	0
X_2	0	0	X_{17}	1	0
X_3	1	1	X_{18}	0	0
X_4	0	0	X_{19}	0	0
X_5	1	1	X_{20}	0	0
X_6	1	0	X_{21}	1	1
X_7	1	1	X_{22}	1	1
X_8	0	0	X_{23}	1	1
X_9	0	0	X_{24}	1	1
X_{10}	1	0	X_{25}	1	1
X_{11}	0	0	X_{26}	0	0
X_{12}	0	0	X_{27}	1	1
X_{13}	0	0	X_{28}	0	0
X_{14}	0	0	X_{29}	1	0
X_{15}	0	0	X_{30}	0	0

Table 3.28 Distances between final cluster centers—government agencies in National

Cluster	1	2
1		4.688
2	4.688	

Table 3.29 Number of cases in each cluster—government agencies in National

Cluster	1	3.000
	2	17.000
Valid		20.000
Missing		0

finally attributed in a variable Xi (Table 3.30). Additionally, a 2-dimentional table was developed and was used in order to examine the existence of the criteria and evaluate the policies in the websites. For that purpose, the values were attributed to variables X1 to X6, respectively.

K-means analysis was used to identify homogenous groups of the prefectures according to their agricultural exploitations. The initial cluster centers are presented in Table 3.31. They are vectors with their values based on the 8 variables, which refer to the two clusters. These 2 clusters are at maximum index distance from each other.

In Table 3.32 we can see the number of the iterations and the changes in the cluster centers. In the 15th iteration the process of redistribution of the units stops because there are no changes in the cluster centers.

The results are summarized in Table 3.33, i.e. which cluster each unit belongs to and the new cluster centers. The first cluster is formed by the cases P_1, P_3, P_4,

Table 3.30 Variables and data—agricultural exploitations

Variables Prefecture/Cases	X_1 Number of municipalities	X_2 Total area (thousand stremmas)	X_3 Cultivated and fallow areas (thousand stremmas)	X_4 Pasture (thousand stremmas)	X_5 Forests (thousand stremmas)	X_6 Areas covered by water (thousand stremmas)	X_7 Settlements (thousand stremmas)	X_8 Other areas (thousand stremmas)
P_1	24	3272.7	1313.9	121.8	1536.7	14.7	75.9	209.7
P_2	22	2622.5	1699.8	48.9	730.7	39.2	54.9	49
P_3	29	5423	1725.2	395.6	2765.1	278.7	36.2	222.2
P_4	20	2953.3	1304	298.7	1256.3	27.4	31.4	35.4
P_5	27	4164.2	1404.8	302.2	2236.4	16.9	34.4	169.7
P_6	11	1870.6	148.2	47.7	1457.6	54.6	2.5	160
P_7	25	4439.9	1896.4	291.7	2144.4	24	49.9	33.5
P_8	12	2126.2	312.6	324.8	1347.6	18.3	9.3	113.6
P_9	16	1609.8	480.9	85.4	835.6	74.8	9.4	123.7
P_10	10	1517.8	368.7	210.2	758.5	18.5	5.4	156.6
P_11	41	4998.9	803.9	692.4	3206.7	58.9	35.3	201.8
P_12	9	1036.3	394.5	176.6	404	19.4	8.6	33.3
P_13	21	2638	1330.9	133.3	991.6	28.2	51	103
P_14	31	5385.6	2778.8	693	1766.3	21.1	91.8	34.6
P_15	26	2636.7	1042.2	144	1378.6	5.4	55.5	11
P_16	26	3386.1	839.8	590.6	1727.4	27.1	38.2	163
P_17	9	3468.8	733.4	313.6	2203.4	18.5	33.1	166.8
P_18	11	2117.3	725.7	188.3	1037.4	45.4	33.7	86.8
P_19	13	4248	2217.7	182.4	1638.7	129	56.1	24.2

(continued)

3.3 K-Means Analysis

Table 3.30 (continued)

Variables Prefecture/Cases	X_1 Number of municipalities	X_2 Total area (thousand stremmas)	X_3 Cultivated and fallow areas (thousand stremmas)	X_4 Pasture (thousand stremmas)	X_5 Forests (thousand stremmas)	X_6 Areas covered by water (thousand stremmas)	X_7 Settlements (thousand stremmas)	X_8 Other areas (thousand stremmas)
P_20	10	1795.6	578.2	119	966.5	75.1	21.9	34.9
P_21	12	2550.2	1077.3	135	1114.4	82.4	35.3	105.8
P_22	12	1703.5	857.4	104.7	663.1	29.9	45.1	3.4
P_23	45	3680.9	1973.1	254.9	1102.6	153.2	152.1	45
P_24	12	2524.4	1326.5	395	715.8	30.8	33.4	22.9
P_25	11	2505.8	1029.1	266.2	1065	37.9	46.3	61.4
P_26	13	1523.9	686.8	77.8	688.4	21.9	33.5	15.4
P_27	27	3970.9	2061.5	285.9	1359.1	105.4	77.3	81.7
P_28	15	3260.6	1399.7	53.3	1720.3	9	20.7	57.6
P_29	15	2296.4	686.7	266.4	1305.3	2	11.9	24
P_30	15	1724.3	414.2	320.4	919.3	35.7	10.5	24
P_31	19	3518.9	1347.2	705.5	1128.9	62.6	49.7	225
P_32	12	1926.8	643.7	272	806.1	130.3	13	61.7
P_33	6	405.9	210	38.2	133.8	0.5	9.6	13.8
P_34	16	639.9	467.3	30.1	65.3	7.1	31.1	39
P_35	9	902.4	259.2	140.4	390.1	0.5	16.5	95.7
P_36	8	354.9	167.2	30.3	122.4	1.5	6.9	26.6
P_37	16	2154.6	863.4	244.6	936	3.6	26.4	80.6
P_38	23	4418.4	1321.8	370.9	2525.5	10.6	38.9	150.7

(continued)

Table 3.30 (continued)

Variables Prefecture/Cases	X_1 Number of municipalities	X_2 Total area (thousand stremmas)	X_3 Cultivated and fallow areas (thousand stremmas)	X_4 Pasture (thousand stremmas)	X_5 Forests (thousand stremmas)	X_6 Areas covered by water (thousand stremmas)	X_7 Settlements (thousand stremmas)	X_8 Other areas (thousand stremmas)
P_39	15	2296.2	1025.1	77.5	980	6.2	51.7	155.7
P_40	22	3639.5	1367.5	204.6	1842.5	7.8	37.6	179.5
P_41	31	2996.9	1614.5	55.9	1189	8.7	46.4	82.4
P_42	18	2151.4	894.5	525.1	643.9	21.1	25.4	41.4
P_43	8	780.6	251.2	55.2	428	0.6	8.8	36.8
P_44	10	907.2	254.9	289.5	287.5	0.1	9.7	65.5
P_45	27	2717.1	765.9	520.4	1200.5	13.1	52.3	164.9
P_46	31	2599.4	836.1	789.5	690.4	3	33.7	246.7
P_47	26	2640.6	1533.4	721	272.6	0.7	46.9	66
P_48	8	1827.2	661.1	624.1	412.3	1.3	11.1	117.4
P_49	11	1495.6	626.8	516.6	285.6	0.1	14.1	52.4
P_50	25	2349.5	765.6	498.9	882.3	0.7	34.1	167.9

3.3 K-Means Analysis

Table 3.31 Initial cluster centers—agricultural exploitations

Variables	Cluster	
	1	2
X_1	29	8
X_2	5423.0	354.9
X_3	1725.2	167.2
X_4	395.6	30.3
X_5	2765.1	122.4
X_6	278.7	1.5
X_7	36.2	6.9
X_8	222.2	26.6

Table 3.32 Iteration history—agricultural exploitations

Iteration	Change in cluster centers	
	1	2
1	1757.974	1643.689
2	70.526	120.753
3	4.149	3.450
4	0.244	0.099
5	0.014	0.003
6	0.001	8.047E-5
7	4.967E-5	2.299E-6
8	2.922E-6	6.569E-8
9	1.719E-7	1.877E-9
10	1.011E-8	5.360E-11
11	5.954E-10	1.462E-12
12	3.515E-11	0
13	1.714E-12	0
14	6.355E-14	0
15	0	0

P_5, P_7, P_11, P_14, P_16, P_17, P_19, P_23, P_27, P_28, P_31, P_38, P_40, P_41 and the second by P_2, P_6, P_8, P_9, P_10, P_12, P_13, P_15, P_18, P_20, P_21, P_22, P_24, P_25, P_26, P_29, P_30, P_32, P_33, P_34, P_35, P_36, P_37, P_39, P_42, P_43, P_44, P_45, P_46, P_47, P_48, P_49, P_50.

In Table 3.34 we can see the final cluster centers, and in Table 3.35—the distance between the final cluster centers.

Table 3.36 presents data for the number of units in each cluster as well as their total number and missing units (if there are any).

Table 3.33 Cluster membership—agricultural exploitations

Prefecture	Cluster	Distance	Prefecture	Cluster	Distance
P_1	1	887.134	P_26	2	412.624
P_2	2	1243.579	P_27	1	737.732
P_3	1	1687.993	P_28	1	845.108
P_4	1	1262.692	P_29	2	682.445
P_5	1	418.202	P_30	2	388.536
P_6	2	926.934	P_31	1	997.184
P_7	1	617.466	P_32	2	148.885
P_8	2	756.428	P_33	2	1701.388
P_9	2	414.084	P_34	2	1467.221
P_10	2	517.458	P_35	2	1153.858
P_11	1	1845.410	P_36	2	1763.621
P_12	2	981.480	P_37	2	350.377
P_13	2	1003.401	P_38	1	795.328
P_14	1	1878.552	P_39	2	588.643
P_15	2	1030.630	P_40	1	450.031
P_16	1	994.487	P_41	1	1269.501
P_17	1	1037.441	P_42	2	437.731
P_18	2	365.277	P_43	2	1262.083
P_19	1	774.570	P_44	2	1184.098
P_20	2	301.321	P_45	2	983.490
P_21	2	843.191	P_46	2	922.359
P_22	2	297.128	P_47	2	1302.036
P_23	1	965.071	P_48	2	522.015
P_24	2	895.239	P_49	2	678.766
P_25	2	756.401	P_50	2	550.543

Table 3.34 Final cluster centers—agricultural exploitations

Variables	Cluster	
	1	2
X_1	25	15
X_2	3954.5	1846.9
X_3	1535.5	708.6
X_4	341.9	255.1
X_5	1844.1	755.0
X_6	57.3	24.5
X_7	53.2	26.0
X_8	122.5	77.6

Table 3.35 Distances between final cluster centers—agricultural exploitations

Cluster	1	2
1		2514.568
2	2514.568	

Table 3.36 Number of cases in each cluster—agricultural exploitations

Cluster	1	17
	2	33
Valid		50
Missing		0

3.3.3 Classification of Agrifood Entities

The websites of 20 agrifood enterprises were used for this example. Initially, qualitative analysis was performed in order to examine the type of management and marketing criteria found in these agrifood enterprises websites; then a quantitative analysis was carried out, in order to examine the presence or absence of these criteria/characteristics (Tsekouropoulos et al. 2012a).

Various management and marketing criteria were introduced in the retrieved corporate websites and different criteria were identified and introduced in each website. Each management and marketing function constitutes a criterion and it is finally attributed in a variable X_i (Table 3.37). Additionally, a 2-dimentional table was developed and was used in order to examine the existence of the criteria and evaluate the policies in the websites. For that purpose, the values were attributed to variables X_1 to X_6, respectively.

Variable X_1 deals with programming of the agrifood enterprise. Programming is a primary function of management and deals with the determination of the goals for the enterprise. It contains all the activities that are anticipated and defined. Variable X_2 refers to the organizational structure of the enterprise. The organizational structure of an enterprise concerns the way that its organization is formed, also the division of the several tasks they exist and the posts of work. Organogrammes (charts) are diagrammes that show the form of the responsibilities in an organization and can be done for the entire enterprise or just a part of it. The length depends on the enterprise size. Variable X_3 stands for the systems of guarantying quality of the enterprise. Systems of guarantying quality include all the programmed and systematic actions that are necessary in order to provide warranty that a product or service will meet all the required needs of quality. Variable X_4 refers to the online advertisement. On line advertisement is the kind of advertisement that exclusively uses the internet as a means of communication and promotion. Variable X_5 deals with the CSR activities of the agrifood enterprises. Variable X_6 stands for the programs of public relationships and generally speaking communication policy of

Table 3.37 Variables attributed to criteria to be achieved by the agrifood entities

Variable	Function	Description of web service
X_1	Management	Management operations planning (Programming)
X_2		Organogram—Organizational structure
X_3		Systems of guarantying quality
X_4	Marketing	On line advertisement
X_5		CSR activities
X_6		Programs of public relationships and generally speaking communication policy

Table 3.38 Data—agrifood entities

A_E	X_1	X_2	X_3	X_4	X_5	X_6
A_E_1	1	1	1	1	1	1
A_E_2	1	1	1	1	0	1
A_E_3	1	0	0	1	0	1
A_E_4	0	1	1	1	0	0
A_E_5	1	1	1	1	0	1
A_E_6	1	1	1	1	1	1
A_E_7	0	1	1	1	0	0
A_E_8	1	0	1	1	0	0
A_E_9	1	1	1	1	1	1
A_E_10	1	1	1	1	0	1
A_E_11	1	1	1	1	1	1
A_E_12	1	1	1	1	0	1
A_E_13	1	1	1	1	1	1
A_E_14	1	1	1	1	1	1
A_E_15	1	1	1	1	0	0
A_E_16	1	1	1	1	1	0
A_E_17	1	1	1	1	1	1
A_E_18	0	1	1	1	0	0
A_E_19	1	1	1	1	1	0
A_E_20	1	1	1	1	0	0

enterprises, through which an enterprise approaches and communicates its buyers, partners and personnel.

Whenever a criterion was achieved for a website the value 1 was attributed to the respective variable aiming at justifying the relative service within the evaluation of the website. The findings are presented in Table 3.38 and it was further analyzed the achievement of each service criteria in the sample websites.

K-means analysis was used to identify homogenous groups of agrifood entities that have similar management and marketing criteria characteristics but are

3.3 K-Means Analysis

Table 3.39 Initial cluster centers—agrifood entities

Variables	Cluster	
	1	2
X_1	0	1
X_2	1	0
X_3	1	0
X_4	1	1
X_5	0	0
X_6	0	1

Table 3.40 Iteration history—agrifood entities

Iteration	Change in cluster centers	
	1	2
1	1.128	0
2	0	0

distinctively different from other agrifood entities segments. The initial cluster centers are presented in Table 3.39. They are vectors with their values based on the 6 variables, which refer to the two clusters. These 2 clusters are at maximum index distance from each other.

In Table 3.40 we can see the number of the iterations and the changes in the cluster centers. In the second iteration the process of redistribution of the units stops because there are no changes in the cluster centers.

The results are summarized in Table 3.41, i.e. which cluster each unit belongs to and the new cluster centers. The first cluster is formed by the cases A_E_1, A_E_2, A_E_4, A_E_5, A_E_6, A_E_7, A_E_8, A_E_9, A_E_10, A_E_11, A_E_12, A_E_13, A_E_14, A_E_15, A_E_16, A_E_17, A_E_18, A_E_19, A_E_20 and the second by A_E_3.

In Table 3.42 we can see the final cluster centers, and in Table 3.43—the distance between the final cluster centers.

Table 3.44 presents data for the number of units in each cluster as well as their total number and missing units (if there are any).

3.4 Combining PROMETHEE II and Clustering for Decision Making

The combination of PROMETHEE II method and clustering was described and presented in detail in the second chapter and it was successfully applied in assessment of ICT adoption, agricultural and environmental topics such as ranking and clustering of aquaculture units (Andreopoulou et al. 2009).

Table 3.41 Cluster membership—agrifood entities

Case number	Cluster	Distance
A_E_1	1	.694
A_E_2	1	.655
A_E_3	2	.000
A_E_4	1	1.128
A_E_5	1	.655
A_E_6	1	.694
A_E_7	1	1.128
A_E_8	1	1.217
A_E_9	1	.694
A_E_10	1	.655
A_E_11	1	.694
A_E_12	1	.655
A_E_13	1	.694
A_E_14	1	.694
A_E_15	1	.766
A_E_16	1	.800
A_E_17	1	.694
A_E_18	1	1.128
A_E_19	1	.800
A_E_20	1	.766

Table 3.42 Final cluster centers—agrifood entities

Variables	Cluster	
	1	2
X_1	1	1
X_2	1	0
X_3	1	0
X_4	1	1
X_5	0	0
X_6	1	1

Table 3.43 Distances between final cluster centers—agrifood entities

Cluster	1	2
1		1.524
2	1.524	

3.4.1 Ranking and Classification of Enterprises that Promote Nature Activities in National Parks

The websites of 20 enterprises that promote nature activities in the 10 Greek National Parks were used for this example. Initially, qualitative analysis was performed in order to examine the type of common e-services criteria found in these

3.4 Combining PROMETHEE II and Clustering for Decision Making

Table 3.44 Number of cases in each cluster—agrifood entities

Cluster	1	19
	2	1
Valid		20
Missing		0

Table 3.45 Variables attributed to criteria to be achieved by the government agencies in National Parks

Variable	Feature	Variable	Feature
X_1	Two or more languages	X_{16}	Links to other companies etc.
X_2	Information about products-services-activities	X_{17}	Various topics of interest
X_3	Contact information	X_{18}	Downloadable files
X_4	Local information	X_{19}	Calendar application
X_5	Digital map	X_{20}	Event calendar application
X_6	Audiovisual material	X_{21}	Celebration calendar application
X_7	Live web camera	X_{22}	Social media sharing
X_8	Search engine	X_{23}	Social media profile
X_9	Sitemap	X_{24}	Forum
X_{10}	Updated enterprise information	X_{25}	Related sources of information
X_{11}	Online survey	X_{26}	Third person advertisement
X_{12}	Online communication form	X_{27}	Newsletter
X_{13}	Weather forecast	X_{28}	RSS
X_{14}	Website visitor tracker	X_{29}	Code access
X_{15}	Frequently Asked Questions (FAQ)	X_{30}	Personalization of the page, trace, safety

enterprises websites; then a quantitative analysis was carried out, in order to examine the presence or absence of these criteria/characteristics.

Various e-services were introduced in the retrieved websites and 30 different criteria were identified and introduced in each website. Each e-service constitutes a criteria/characteristic and it is finally attributed in a variable Xi (Table 3.45). Additionally, a 2-dimentional table was developed and was used in order to examine the existence of the criteria and evaluate the services of the websites. For that purpose, the values were attributed to variables X_1 to X_{30}, respectively.

The first variable refers to the ability to view the content of an internet presence in more than two languages (Greek and English). Variable X_2 refers to the provision of detailed information about the products or services of the enterprise and the third one refers to the provision of information about the enterprise or organization (ownership, mailing address, telephone number and email address) that facilitate the communication between the customers and the enterprises. As for the variable X_4, it refers to the provision of information about the local area.

Variable X_5 is associated with the provision of an interactive digital map for the orientation of website visitors–customers. The sixth variable refers to the provision of any audiovisual material, such as photographs, videos and virtual tours, while the seventh variable refers to the existence of live web camera. Variable X_8 represents the provision of web search engine. Variable X_9 refers to the provision of a sitemap for an overview of the website content. The tenth variable is associated with the continuous updating of the internet presence related to the activities of the enterprise. As for the variable X_{11}, it refers to the provision of the application for online polls to identify the users' trends. Variable X_{12} refers to the provision of online communication form. Variable X_{13} is associated with the provision of weather forecast application, while variable X_{14} refers to the website visitor tracker. Variable X_{15} refers to the tab of the Frequently Asked Questions (FAQ) and variable X_{16} is associated with the provision of useful links to other relevant organizations or enterprises. Variable X_{17} refers to the provision of information on different topics, while variable X_{18} refers to the ability to download some useful files. Variable X_{19} is associated with the clock and calendar application, variable X_{20} is associated with the local events calendar and variable X_{21} with the celebration calendar application integration. Variable X_{22} refers to the provision of sharing the internet presence through users' account in social networks, while variable X_{23} refers to the participation of the enterprise in social media. Variable X_{24} is associated with the provision of an online interactive community through the internet presence and variable X_{25} is associated with the available information on related topics. Variable X_{26} refers to the third person advertisement. As for the variable X_{27}, it refers to the provision of newsletter service. Variable X_{29} refers to the ability to create a user account. Variable X_{30} refers to the ability to be personalized the website by the registered users, while variable X_{28} refers to the RSS service, which distributes information through the Internet. The subscription to a website RSS removes the need of manually checking the website for new content because the users' browser regularly monitors the internet presence and informs the users of any updates.

Whenever a criterion was achieved for a website the value 1 was attributed to the respective variable aiming at justifying the relative service within the evaluation of the website. The findings are presented in Table 3.46 and it was further analyzed the achievement of each service criteria in the sample websites.

Then, the total ranking of the websites was studied. The method that was used for the total ranking was the multicriteria analysis named PROMETHEE II. That method applies a linear form of service in this particular case, using the e-services of the websites identified as criteria.

The net flow is the final number that is used for the comparison between the websites in order to obtain the ranking. The ten values (scenarios) range between 0.25 and 2.5s with step 0.25s, where s is the standard deviation of all differences d for each criterion. In total, we take 500 net flow values for each website and find the website's average value. Each website with a higher net flow is considered superior in the final ranking.

3.4 Combining PROMETHEE II and Clustering for Decision Making

Table 3.46 Data—enterprises that promote nature activities in National Parks

N_A	X_1	X_2	X_3	X_4	X_5	X_6	...	X_{25}	X_{26}	X_{27}	X_{28}	X_{29}	X_{30}
N_A_1	3	1	1	1	0	1	...	0	0	0	0	0	0
N_A_2	2	1	1	1	0	1	...	0	1	0	0	0	0
N_A_3	7	1	1	1	0	1	...	1	1	0	0	0	0
N_A_4	1	1	1	1	0	1	...	1	1	1	0	1	1
N_A_5	1	1	1	1	1	1	...	0	1	1	0	1	1
N_A_6	1	1	1	1	1	1	...	0	0	0	0	0	0
N_A_7	1	1	1	1	0	1	...	1	1	0	0	1	1
N_A_8	2	1	1	1	0	1	...	1	0	1	0	1	1
N_A_9	1	1	1	1	1	1	...	1	1	1	0	0	0
N_A_10	1	1	1	1	0	1	...	0	0	0	0	0	0
N_A_11	2	1	1	1	1	1	...	1	0	0	0	0	0
N_A_12	1	1	1	1	0	1	...	0	0	0	0	0	0
N_A_13	3	1	1	1	0	1	...	0	0	0	0	0	0
N_A_14	1	1	1	0	0	1	...	0	0	0	0	0	0
N_A_15	1	1	1	1	0	1	...	0	0	0	0	0	0
N_A_16	1	1	1	0	0	1	...	0	0	0	0	0	0
N_A_17	4	1	1	1	0	1	...	0	1	0	0	0	0
N_A_18	2	1	1	1	0	1	...	1	0	0	0	0	0
N_A_19	2	1	1	1	1	1	...	1	0	0	0	0	0
N_A_20	2	1	1	1	1	1	...	1	1	0	0	0	0

According to these findings (Table 3.47), the values estimated for total net flows ϕ present a great spectrum of values between +5,26 to −2,85 and that indicates a great difference concerning "superiority" between the first and the last case in the ranking of the websites. The higher the net flow is, the better that respective alternative is (Euroconsultants 2011). As a result of the calculations made, we obtained the following ranking of the alternatives/cases studied: "N_A_19" alternative ranked the first, "N_A_18" alternative ranked the second, "N_A_8" alternative ranked the third, "N_A_10" alternative ranked the fourth, etc. From the study made, one notices that "N_A_19" alternative ranked the first, and therefore we recommend the "N_A_19" as the best solution. The enterprises that promote nature activities in National Parks should definitely evolve and include innovative e-service features to improve their websites aiming to promote their activities and the website of "N_A_19" should be their model in that process.

Furthermore, cluster analysis is used to facilitate the decision-making process. The use of multicriteria analysis for the total ranking of the enterprises websites and the performance of cluster analysis for their classification in clusters of common content characteristics are important within the target for assessing the possibilities of an effective exploitation of these e-services by the enterprises (Arabatazis et al. 2010).

Table 3.47 Total net flows—enterprises that promote nature activities in National Parks

	N_A	Net flow—linear
1	N_A_19	5.26702991
2	N_A_18	3.185655462
3	N_A_8	3.02815776
4	N_A_10	2.704851267
5	N_A_7	2.125487862
6	N_A_16	1.889908313
7	N_A_5	1.714004849
8	N_A_12	1.518905271
9	N_A_3	1.367911741
10	N_A_17	1.194371661
11	N_A_4	1.133337259
12	N_A_1	0.842077978
13	N_A_14	0.579054379
14	N_A_11	0.518300458
15	N_A_2	0.491513409
16	N_A_9	0.470360902
17	N_A_6	0.217370649
18	N_A_15	−0.664809423
19	N_A_13	−0.721499514
20	N_A_20	−2.857599504

The e-services characteristics were classified in four groups, each one representing a stage of e-services adoption (Koliouska and Andreopoulou 2013). The four groups are the following: 1. presence, 2. interaction, 3. transaction and 4. transformation. "Presence" is called the primary stage which ensures that the website is accessible for all users in many ways and the users interact with the website interface in order to acquire information. The e-services characteristics X_1, X_2, X_3 and X_4 are classified in the first stage of "presence". "Interaction" is the second stage of ICT adoption that enables the website visitors to accomplish whatever process or experience is offered by the website. Hence, in this stage of interaction there are few actions enabled for the visitor, such as the navigation to the website and the provision of useful links. So, the e-services characteristics X_5 to X_{21} belong to the second stage of "interaction". "Transaction" addresses the e-shopping experience by the purchasing process and the payment orders. In this stage of "transaction" there are implemented applications for transactions where the user plays a main role, such as electronic exchange of self-services and texts provided for the user. The e-services characteristics X_{22} to X_{26} belong to the third stage of "transaction". "Transformation" is related with the quality of communication and transaction experience in addition to the responsiveness and reliability to the customers. Thus, in this stage of "transformation" the value chain gets improved while the users have the possibility of making online orders and payments while at the same time, they can check the stage of their order. Finally, the rest e-services

Table 3.48 Variable t—enterprises that promote nature activities in National Parks

N_A	t_1	t_2	t_3	t_4
N_A_1	4	3	1	0
N_A_2	4	5	1	0
N_A_3	4	9	4	0
N_A_4	3	5	2	3
N_A_5	3	8	3	3
N_A_6	3	6	1	0
N_A_7	3	4	5	2
N_A_8	4	6	2	3
N_A_9	3	10	5	1
N_A_10	3	4	1	0
N_A_11	4	6	1	0
N_A_12	3	3	2	0
N_A_13	4	6	0	0
N_A_14	2	2	0	0
N_A_15	3	4	1	0
N_A_16	2	2	2	0
N_A_17	4	3	1	0
N_A_18	4	2	1	0
N_A_19	4	4	2	0
N_A_20	4	8	4	0

characteristics X_{27} to X_{30} represent the final stage of "transformation" (Andreopoulou et al. 2007a; Andreopoulou et al. 2008).

The example also included the total amount of criteria achieved of each enterprise website in each e-services adoption stage. The number of present e-services characteristics are attributed to a new variable, named T. Variable T presents the number of characteristics/achieved criteria that are achieved in each enterprise website where T is a number between 1 and n for each group (e.g. $1 \leq t_1 \leq 4, 1 \leq t_2 \leq 17, 1 \leq t_3 \leq 5$ and $1 \leq t_4 \leq 4$) (Table 3.48).

Hierarchical cluster analysis was used to identify homogenous groups of enterprises that promote nature activities in National Parks and have similar e-services characteristics but are distinctively different from other enterprise segments. In Table 3.49, the case processing summary is presented which lists the number of valid cases and the number of missing cases.

The proximity matrix (Table 3.50) provides the actual distances, which reveals the similarities computed for any pair of observations and variables.

Table 3.49 Case processing summary—wood enterprises

Cases					
Valid		Missing		Total	
N	Percent	N	Percent	N	Percent
20	100.0	0	0	20	100.0

Table 3.50 Proximity matrix—enterprises that promote nature activities in National Parks (part)

Case	Squared euclidean distance								
	1	2	3	4	5	6	7	8	9
1	0	4.00	45.00	15.00	39.00	10.00	22.00	19.00	67.00
2	4.00	0	25.00	11.00	23.00	2.00	22.00	11.00	43.00
3	45.00	25.00	0	30.00	12.00	19.00	31.00	22.00	4.00
4	15.00	11.00	30.00	0	10.00	11.00	11.00	2.00	38.00
5	39.00	23.00	12.00	10.00	0	17.00	21.00	6.00	12.00
6	10.00	2.00	19.00	11.00	17.00	0	24.00	11.00	33.00
7	22.00	22.00	31.00	11.00	21.00	24.00	0	15.00	37.00
8	19.00	11.00	22.00	2.00	6.00	11.00	15.00	0	30.00
9	67.00	43.00	4.00	38.00	12.00	33.00	37.00	30.00	0

Table 3.51 Agglomeration schedule—enterprises that promote nature activities in National Parks

Stage	Cluster combined		Coefficients	Stage cluster first appears		Next stage
	Cluster 1	Cluster 2		Cluster 1	Cluster 2	
1	1	17	0	0	0	4
2	10	15	0	0	0	8
3	3	20	1.000	0	0	13
4	1	18	1.000	1	0	11
5	11	13	1.000	0	0	6
6	2	11	1.500	0	5	7
7	2	6	1.667	6	0	16
8	10	19	2.000	2	0	11
9	12	16	2.000	0	0	12
10	4	8	2.000	0	0	15
11	1	10	3.000	4	8	12
12	1	12	4.167	11	9	14
13	3	9	5.500	3	0	19
14	1	14	6.375	12	0	16
15	4	5	8.000	10	0	17
16	1	2	10.306	14	7	18
17	4	7	15.667	15	0	18
18	1	4	22.019	16	17	19
19	1	3	37.510	18	13	0

The agglomeration schedule (Table 3.51) details how observations are clustered together at each stage of the cluster analysis. When clusters or cases are joined, they are subsequently labeled with the smaller of the two cluster numbers. The Coefficients column indicates the distance between the two clusters (or cases) joined at each stage. For a good cluster solution, you will see a sudden jump in the

3.4 Combining PROMETHEE II and Clustering for Decision Making

Table 3.52 Cluster membership—enterprises that promote nature activities in National Parks

N_A	2 Clusters	N_A	2 Clusters
N_A_1	1	N_A_11	1
N_A_2	1	N_A_12	1
N_A_3	2	N_A_13	1
N_A_4	1	N_A_14	1
N_A_5	1	N_A_15	1
N_A_6	1	N_A_16	1
N_A_7	1	N_A_17	1
N_A_8	1	N_A_18	1
N_A_9	2	N_A_19	1
N_A_10	1	N_A_20	2

distance coefficient (or a sudden drop in the similarity coefficient) as you read down the table (Stage 18: Coefficient 22,019, Stage 19: Coefficient 37,510). The stage before the sudden change indicates the optimal stopping point for merging clusters (Stage 18). For this example, 2 clusters remain after Stage 18. So, we should consider using a 2-cluster solution.

The cluster membership (Table 3.52) provides detailed group structure after classification.

This plot (Fig. 3.5) gives a graphic representation of how the cases are joined at each stage of the analysis. Each white bar represents a boundary between clusters. At each stage, two clusters are joined, and so the white bar separating the joined clusters ends. Within a row, each contiguous black band indicates cases grouped as a cluster.

The results of the cluster analysis is a dendrogram (Fig. 3.6) with $19(n-1)$ nodes. The horizontal axis of the dendrogram represents the distance or dissimilarity between clusters. The vertical axis represents the cases and clusters. Each joining (fusion) of two clusters is represented on the graph by the splitting of a horizontal line into two horizontal lines. The horizontal position of the split, shown by the short vertical bar, gives the distance between the two clusters. As longer the horizontal line is, more stable the previous clustering tends to be.

The number of clusters the will be formed at a particular Cluster Cutoff value may be quickly determined from this plot by drawing a vertical line at that value and counting the number of lines that the vertical line intersects. For example, we can see that if we draw a vertical line at the value 1, four clusters will result. One cluster will contain thirteen objects, one cluster will contain four objects and one cluster will contain three objects. Cluster analysis consists a tool to create a typology for websites characteristics of wood enterprises.

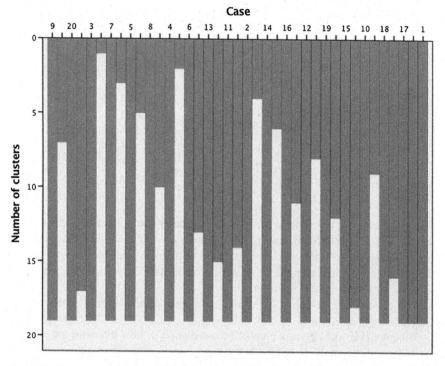

Fig. 3.5 Clustering with "Between Groups Linkage method"—wood enterprises

3.4.2 Ranking and Classification of Enterprises that Promote Rural Production

The websites of 20 enterprises that promote rural production were used for this example. Initially, qualitative analysis was performed in order to examine the type of common e-services criteria found in these enterprises websites; then a quantitative analysis was carried out, in order to examine the presence or absence of these criteria/characteristics (Andreopoulou et al. 2011).

Various e-services were introduced in the retrieved websites and 13 different criteria were identified and introduced in each website. Each e-service constitutes a criteria/characteristic and it is finally attributed in a variable Xi (Table 3.53). Additionally, a 2-dimentional table was developed and was used in order to examine the existence of the criteria and evaluate the services of the websites. For that purpose, the values were attributed to variables X_1 to X_{13}, respectively.

Variable X_1 refers to the detailed information on the agrotourism enterprise, the supported activities and services. Variable X_2 represents the existence of the current price list for the offered services, activities, products and special offers. Variable X_3 stands for the existence of contact information in the type of telephone, address, e-mail, etc. to enhance communication with the enterprise. Variable X_4 refers to

3.4 Combining PROMETHEE II and Clustering for Decision Making

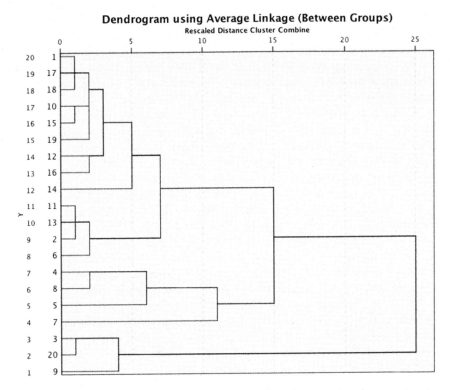

Fig. 3.6 Dendrogram using Average Linkage (Between Groups)—enterprises that promote nature activities in National Parks

Table 3.53 Variables attributed to criteria to be achieved by the enterprises that promote rural production

Criteria	Main features of each
X_1	Information about products-services-activities
X_2	Current prices
X_3	Contact Information
X_4	Local information
X_5	Links to other companies etc.
X_6	Related sources of information
X_7	On line reservation (enabled with traditional ways of payment)
X_8	On line reservation (enabled with online payment)
X_9	On line communities (forums, chat rooms, guestbooks etc.)
X_{10}	Additional topics with information on different categories of interest
X_{11}	Code access
X_{12}	Third person advertisement
X_{13}	Personalization of the page, trace, safety

features such as "local information" usually with texts, photo galleries and maps of the local area with tourism attractions and landmarks aiming to appeal the potential visitor. Variable X_5 represents the provided links to other similar entities, usually locally. Variable X_6 stands for information for other related sources of information concerning agrotourism, sustainable development, sustainable agriculture, development programs through the EU, such as LEADER+, local development agencies, etc. Variable X_7 refers to the feature about online reservation where website users can easily proceed and reserve their staying in the agotourism enterprise through reservation forms. Variable X_8 represents the online payment feature. Variable X_9 stands for the communication features for the users of the website such as forums, guestbook etc. where visitors can share their experience and communicate. Variable X_{11} stands for the existence of code access involving membership, while variable X_{12} stands for the third person advertisement. Finally, variable X_{13} refers to the personalization of the website through safety features. The findings are presented in Table 3.54 and it was further analyzed the achievement of each service criteria in the sample websites. For enterprise Website, the total number of present characteristics is attributed to a new variable; named T. Variable T presents the total of characteristics/achieved criteria 1 to 13 that are present in each Website.

Finally, the optimum group derived from total ranking of the enterprises that promote rural production using the multicriteria analysis PROMETHEE II is presented in Table 3.54. The same Table 3.55 also includes the total net flows estimated for each Website and it is used for the comparison between the Websites in order to obtain the ranking, as each Website with a higher net flow is considered superior in ranking. The total sum of achieved criteria T for each Website is also included in the table. Furthermore, K-means analysis is used to facilitate the decision-making process. The use of multicriteria analysis for the total ranking of the enterprises websites and the performance of cluster analysis for their classification in 3 clusters of common variable t and total net flow (φ) are important within the target for assessing the possibilities of an effective exploitation of these e-services by the enterprises.

According to these findings, the values estimated for total net flows ϕ present a great spectrum of values between +6,476 to −3,337 and indicate a great difference concerning "superiority" between the first and the last case in the ranking of the enterprises' Websites. Moreover, the total flows ϕ of the enterprises' Websites, as derived from the application of PROMETHEE II method, allow a further grouping of the cases and to generate 3 groups as described in the following section.

Group-1: The "Optimum Group"
In this group, seven enterprises that promote rural production are classified that achieve 7–11 criteria and they present a very high total flows (6,476–1,357) that present a "high superiority" against the rest.

Group-2: The "Equilibrium Group"
In this group, six enterprises that promote rural production have been classified, that achieve 5–6 criteria and close to zero total flows (−0,739–0,441) that present an "equilibrium between superiority and lag" against the rest of the cases (Table 3.56).

3.4 Combining PROMETHEE II and Clustering for Decision Making

Table 3.54 Data—enterprises that promote rural production

	X_1	X_2	X_3	X_4	X_5	X_6	X_7	X_8	X_9	X_{10}	X_{11}	X_{12}	X_{13}	t
R_P_1	1	1	1	1	1	0	1	0	1	1	1	1	1	11
R_P_2	1	1	1	1	0	0	0	0	0	0	0	0	0	4
R_P_3	1	0	1	1	1	0	0	0	0	1	0	1	0	6
R_P_4	1	1	1	1	0	0	0	1	0	0	0	0	0	5
R_P_5	1	0	1	1	1	0	0	0	1	1	0	1	0	7
R_P_6	1	1	1	1	1	0	0	1	0	1	0	1	1	9
R_P_7	1	1	1	1	0	0	0	1	0	1	0	1	1	8
R_P_8	1	0	1	1	0	0	0	0	0	0	0	0	0	3
R_P_9	1	0	1	1	0	0	0	0	1	1	0	1	0	6
R_P_10	1	0	1	1	0	0	0	0	1	0	0	0	0	5
R_P_11	1	1	1	1	0	0	0	0	0	0	0	0	0	4
R_P_12	1	1	1	1	0	0	0	0	0	0	0	0	0	4
R_P_13	1	0	1	1	1	1	0	0	1	0	0	1	0	7
R_P_14	1	1	1	1	0	1	0	0	0	0	0	0	0	6
R_P_15	1	0	1	1	0	0	0	0	0	0	0	0	0	3
R_P_16	1	0	1	1	0	0	0	0	1	1	0	1	0	6
R_P_17	1	1	1	1	0	0	1	0	1	0	0	0	0	7
R_P_18	1	0	1	0	0	1	0	0	1	0	0	0	1	4
R_P_19	1	1	1	1	1	0	0	0	1	1	0	1	1	9
R_P_20	1	0	1	1	0	0	0	0	0	0	0	0	0	3

Table 3.55 Total ranking of the Websites, sum of achieved criteria for each Website, total net flows and classification, in the optimum group-1

Total ranking	R_P	t	Total net flow
1	R_P_1	11	6.476138954
2	R_P_6	9	4.043754521
3	R_P_19	9	2.047980318
4	R_P_7	8	1.781227343
5	R_P_13	7	1.650627119
6	R_P_17	7	1.527297175
7	R_P_5	7	1.357743442

Table 3.56 Total ranking of the Websites, sum of achieved criteria for each Website, total net flows and classification, in the equilibrium group-2

Total ranking	R_P	t	Total net flow
1	R_P_3	6	0.441608568
2	R_P_14	6	0.437512566
3	R_P_9	6	0.43411159
4	R_P_16	6	−0.303451941
5	R_P_4	5	−0.513457946
6	R_P_10	5	−0.739915931

Table 3.57 Total ranking of the Websites, sum of achieved criteria for each Website, total net flows and classification, in the high lag group-3

Total ranking	R_P	t	Total net flow
1	R_P_2	4	−0.856200904
2	R_P_11	4	−1.112463238
3	R_P_12	4	−1.112463238
4	R_P_18	4	−1.349958255
5	R_P_20	3	−1.349958255
6	R_P_8	3	−1.349958255
7	R_P_15	3	−3.33752219

Group-3: The "High Lag Group"

In that group are classified seven Websites of enterprises that achieve 3-4 criteria and average negative total flows (−0,856) to (−3,337) that present a "high lag" against the rest of the cases (Table 3.57).

3.5 Concluding Remarks and Future Research

The book provided an introduction to the operational research, focused on the need for analysis and the ways of use of research results, presented the most common multicriteria ranking technique, PROMETHEE II method and two of the most common cluster generation methods, hierarchical cluster analysis and K-means analysis, are explored, and finally, presented and discussed ten examples in order to

3.5 Concluding Remarks and Future Research

understand how operational research is used in the agricultural, food and environment sector.

This book constitutes the first handbook for students to cover multicriteria analysis for total ranking and clustering classification techniques and their application in the agri-environmental sector. It is the only book dealing with the use of OR results as it brings together a group of examples on the much debated issue of decision making process of enterprises in the agrifood and environmental sector.

Ranking methods and cluster analysis are suitable for the decision making process in agrifood and environment sector involving in selecting the optimum solutions/cases and generating homogenous clusters. These methods have a number of other advantages. The most important advantage is that they simultaneously integrate more than one criterion into a unique decision making process. Future extend focus on multi objective optimization in decision making. Problems with multiple objectives and criteria are generally known as multiple criteria optimization or multiple criteria decision-making (MCDM) problems. So far, these types of problems have typically been modelled and solved by means of linear programming, while nonlinear is a new approach (Miettinen 2012). Basic concepts and principles of multi-objective optimization (MO) methods and multiple criteria decision making (MCDM) approaches will be the new focus while algorithms are classified based on the role of decision makers (Deb et al. 2016). Deb (2014) states that …"Multi-objective optimization constitutes an integral part within optimization activities while it appears to have a booming practical importance, since almost all real-world optimization problems are ideally suited to be modeled using multiple conflicting objectives".

We are confident that the book will be a useful aid for scientists and decision-makers in the agricultural and environmental sector while reading the book will stimulate a fruitful discussion within scientists and experts and will enhance the employment of the methods as well. We trust the book will advance the methods described in new directions and resolutions with both theoretical and practical insight and applications.

References

Andreopoulou, Z., Koutroumanidis, T. (2009). Assessment of the ICT adoption stage in eco-agrotourim websites in Greece. *6th International Conference of Management of Technological Changes, 3rd–5th September, Alexandroupolis, Greece.* I, pp. 441–444.

Andreopoulou, Z., Arabatzis, G., Manos, B., & Sofios, S. (2007). Promotion of rural regional development through the WWW. *International Journal of Applied Systems Studies., 1*(3), 290–304.

Andreopoulou, Z. S., Tsekouropoulos, G., Koutroumanidis, T., Vlachopoulou, M., & Manos, B. (2008). Typology for e-business activities in the agricultural sector. *International Journal of Business Information Systems, 3*(3), 231–251.

Andreopoulou, Z. S., Kokkinakis, A. K., & Koutroumanidis, T. (2009a). Assessment and optimization of e-commerce websites of fish culture sector. *Operational Research, 9*(3), 293–309.

Andreopoulou, Z. S., Koutroumanidis, T., & Manos, B. (2009b). The adoption of e-commerce for wood enterprises. *International Journal of Business Information Systems., 4*(4), 440–459.

Andreopoulou, Z., Koutroumanidis, T., & Manos, B. (2011). Optimizing collaborative e-commerce websites for rural production using multicriteria analysis. In K. Malik & P. Choudhary (Eds.), *Business organizations and collaborative web: practices, strategies and patterns* (pp. 102–119). USA: IGI Global.

Arabatzis, G., Andreopoulou, Z., Koutroumanidis, T., & Manos, B. (2010). E-government for rural development: classifying and ranking content characteristics of development agencies websites. *Journal of Environmental Protection and Ecology., 11*(3), 1138–1149.

Deb, K. (2014). Multi-objective optimization. In *Search methodologies* (pp. 403–449). US: Springer.

Deb, K., Sindhya, K., Hakanen, J. (2016). Multi-objective optimization. In *Decision Sciences: Theory and Practice* (pp. 145–184). Boca Raton: CRC Press.

Euroconsultants (2011). Raport asupra criteriilor de evaluare a tehnologiilor alternative. Înfiinţarea unei reţele a deşeurilor pentru planificarea durabilă a gestionării deşeurilor solide şi promovarea instrumentelor decizionale integrate în Regiunea Balcani (BALKWASTE). Martie, pp. 13–14.

Koliouska, C., & Andreopoulou, Z. (2013). Assessment of ICT adoption stage for promoting the Greek National Parks. *Procedia Technology, 8,* 97–103.

Koliouska, C., Andreopoulou, Z., Kiomourtzi, F., Manos, B. (2015). E-government for national forest parks in Greece. *7th International Conference on Information and Communication Technologies in Agriculture, Food and Environment (HAICTA 2015), Kavala, Greece, September 17–20.* pp. 117–123.

Miettinen, K. (2012). *Nonlinear multiobjective optimization* (Vol. 12). Heidelberg: Springer.

Tsekouropoulos, G., Andreopoulou, Z., Koliouska, C., Koutroumanidis, T., Batzios, C., & Lefakis, P. (2012a). Marketing policies through the internet: The case of skiing centers in Greece. *Scientific Bulletin-Economic Sciences, 11*(1), 66–78.

Tsekouropoulos, G., Andreopoulou, Z., Koliouska, C., Katsonis, N., Vatis, S.E. (2012a). The role of internet in the promotion of agri-food enterprises: e-marketing, management and organizational functions. *International Conference on Contemporary Marketing Issues (ICCMI 2012), Thessaloniki, Greece, June 13–15*.

CPSIA information can be obtained
at www.ICGtesting.com
Printed in the USA
LVOW01*2322050517
533418LV00009B/142/P